T0269678

Agricultural marketing enterprises for the developing world

with case studies of indigenous private, transnational co-operative and parastatal enterprise

JOHN C. ABBOTT

Former Chief, Marketing and Credit Service,
Food and Agriculture Organization of the United Nations

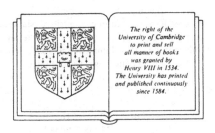

The right of the
University of Cambridge
to print and sell
all manner of books
was granted by
Henry VIII in 1534.
The University has printed
and published continuously
since 1584.

CAMBRIDGE UNIVERSITY PRESS

Cambridge

New York   Port Chester   Melbourne   Sydney

CAMBRIDGE UNIVERSITY PRESS
Cambridge, New York, Melbourne, Madrid, Cape Town, Singapore,
São Paulo, Delhi, Dubai, Tokyo

Cambridge University Press
The Edinburgh Building, Cambridge CB2 8RU, UK

Published in the United States of America by Cambridge University Press, New York

www.cambridge.org
Information on this title: www.cambridge.org/9780521339087

First published 1987
Reprinted 1990
Re-issued in this digitally printed version 2009

A catalogue record for this publication is available from the British Library

Library of Congress Cataloguing in Publication data
Abbott, John Cave, 1919–
Agricultural marketing enterprises for the developing world.
Includes 26 case studies, some published previously,
by 15 authors.
Includes bibliographies.
1. Farm produce – Developing countries – Marketing.
2. Farm produce – Developing countries – Marketing – Case
studies. 3. Produce trade – Developing countries. 4. Produce trade –
Developing countries – Case studies.
I. Title.
HD9018.D44A228   1987   381′.41′091724   86-24462

ISBN 978-0-521-32597-4 Hardback
ISBN 978-0-521-33908-7 Paperback

Contents

Case studies contributions v
Preface vi
Authors and acknowledgements ix
Abbreviations xiii

1 Introduction 1
Marketing: meaning and responsibilities 1
Role of government in marketing 3
Alternative marketing enterprises 5
Issues for discussion 11
Further reading 11

2 The marketing enterprise and economic development 13
The determination of market price 13
Optimum allocation of resources 20
Role of marketing in the development process 21
Marketing for levels of development 22
The marketing enterprise and development 23
The Marxist view 25
Bread and palaces 27
Multiplier effect of the marketing enterprise 28
Issues for discussion 28
Further reading 29

3 Indigenous private enterprise 31
The successful hawker – Dominica 31
Soex, horticultural export enterprise – Senegal 33
Kallu, vegetable producer/wholesaler – Sierra Leone 36
Enebor, rice miller – Nigeria 39
Al Haji, marketing board buying agent – Cameroun 41
Wimalajeewa, market commission agent – Colombo 43
Mafandala, produce buyer and retailer – Zaire 45
Matlhaku, butcher – Botswana 47
The 'market kings' of Sudan 51
Attributes, advantages and support needs 56
Issues for discussion 62

4 **Transnational enterprises** 64
Pepper for Tabasco – Honduras 64
Cassava exports from Thailand 67
Jamaica Broilers 74
Cadbury in India 78
Unilever-Is co-partnership – Turkey 84
Mumias Sugar, Kenya – Booker McConnell 93
Attributes, advantages and support needs 98
Issues for discussion 103
5 **Co-operatives** 105
Ogbomosho Society, marketing food crops – Nigeria 105
Tsaotun Farmers' Association – Taiwan 109
Ha-ee vegetable marketing group – Korea 114
Markfed fertilizer distribution – India 119
Coffee Co-operative Union – West Cameroun 123
Windward Islands Banana Growers (Winban) 128
Attributes, advantages and support needs 132
Issues for discussion 137
6 **Parastatals** 138
BULOG: National food authority – Indonesia 138
Kenya Tea Development Authority 146
Zimbabwe Cotton Marketing Board 152
Botswana Meat Commission 157
Cyprus Potato Marketing Board 165
Attributes, advantages and support needs 172
Issues for discussion 177
7 **Developing an effective marketing structure** 179
Indigenous private enterprise 181
Transnationals 183
Co-operatives 184
Parastatals 185
Constraints and support management 187
Issues for discussion 191
Further reading 191
8 **Managing a marketing enterprise** 192
Planning operations 192
Business management 196
Business controls 204
Assessing performance 208
Issues for discussion 212
Further reading 213
Index 215

Case studies contributed by:

G. Addy, Agricultural Economist, Winban's Research and Development Division, Castries, St Lucia

M. J. Blackie, Professor of Agricultural Economics, University of Zimbabwe, Harare, Zimbabwe

Boonjit Titapinatanakun, Faculty of Economics and Business Administration, Kasetsaart University, Bangkok, Thailand

J. Freivalds, John Freivalds & Associates, 8208 Franklin Street, Minneapolis, Minnesota 55426, U.S.A.

Professor M. Harper, Cranfield School of Management, Bedfordshire

R. Kavura, Institute of Development Management, Tanzania

P. V. John, Senior Manager, Projects and Development India Ltd, New Delhi 110 019, India

Sung Hoon Kim, Professor of Agricultural and Resource Economics, Chung Ang University, Seoul, Korea

L. A. Mears, Food Research Institute, Stanford University, California 94305, USA

H. Mettrick, International Course for Development Orientated Research in Agriculture, Wageningen, The Netherlands

C. A. Osuntogun, Department of Agricultural Economics, University of Ife, Nigeria

M. C. Simpson, School of Economic Studies, University of Leeds, Leeds LS9 9JT

G. A. Truitt, Fund for Multinational Management Research, 684 Park Avenue, New York

Sing Min Yeh, 154 Princeton Street 2, Santa Monica, California 90404, U.S.A.

Preface

A book focussing on enterprises for marketing in the developing world
is badly needed. The bulk of the agricultural marketing literature that
is generally available is concerned with government policy and institutions. The work of the enterprises which provides capital to buy
produce from farmers and capital to store, transport and prepare it for
consumers or manufacturers, has often been neglected. This book will
help to redress the balance. How such enterprises are established and
how they respond to differing sets of conditions is vitally important.
This is the reality of marketing.

For agricultural economic and marketing courses at the intermediate
and degree level this book can be a standard text. An introductory
chapter sets out the functions of agricultural marketing and the main
types of enterprise involved. They are then illustrated in action, with
an analysis of their characteristic attributes, the advantages they offer,
and the support they need from governments and the public if they are
to realize their potential. The penultimate chapter brings this together
in a comparative frame and offers some precepts for policy.

The book features enterprise in development: how marketing enterprises stimulate additional agricultural output, raise farm incomes,
provide services consumers want, earn foreign exchange through
exports, and so contribute to overall development. It will be
appreciated by students of development economics since teaching in
this area has often glossed over the marketing process, taking effective
performance for granted. One chapter focusses specifically on marketing enterprise within the framework of economic development.

Teaching from case studies has for a long time been a favoured
technique in management courses and this book presents 26 studies of
private indigenous, transnational, co-operative and parastatal marketing enterprises. They provide a rich fund of information and experience
on agricultural marketing management. Then follows a chapter which
takes up successive management issues, sets out the guiding prin-

ciples, and then directs the reader to the case studies for illustration and discussion. Texts on agricultural marketing management in the developing countries are rare and this book provides a comprehensive introduction. Teachers of marketing management and business administration with an interest in international affairs will also find this book a valuable complement to texts based on experience in industrialized countries.

General readers will appreciate the objective application of economic principles to agricultural marketing enterprises and techniques. It shows how the performance of individual marketing enterprises can be improved together with the operation of the marketing system of which they form a part. It provides an insight into the continuing relationship between the managers of marketing enterprises and the producers and consumers they serve. The real-life experience presented in the case studies, and the analyses based upon it will find wide appeal. The book will be of particular interest to those concerned with improving marketing in the developing world, to aid agencies seeking to help them, and to farmers and consumers.

Issues for discussion, set out after each chapter, will also serve as exercises for students. It is hoped that readers will be stimulated to become better informed about marketing enterprises and management in their own countries. To assist them, instructors can build up dossiers based on official reports, research studies and newspaper clippings. Students can also be challenged to present and defend ideas on how to improve the performance of some of the enterprises studied or of other enterprises in their own country. They can do this successively as if they were a) managing them, b) responsible for government policy towards them, c) formulating a programme of external assistance.

The enterprises studied have proved themselves over a number of years and have been selected to provide a representative coverage in the developing world. The focus is on how the enterprises faced up to issues calling for decisions. In marketing expertise and technology they range from the continuing use of traditional methods to the relative sophistication of Ha-ee in Korea and Unilever-Is in Turkey. Conditions and requirements are continually changing. These enterprises are to be seen, not so much as examples to follow, as models to build upon and to adapt.

Books for further reading are presented, in a list, not so much to reinforce points made in the text as to provide broader coverage of the subject. Most of them also provide additional references. A lot of information on marketing in developing countries is available in F.A.O. or other aid agency reports and in university research studies in limited circulation. Instructors are urged to obtain photocopies of strategic material and to make them available for students to refer to.

Prices are presented in U.S. dollars converted at the official rate of exchange as it was at the time quoted. Weights and measures are metric.

Authors and acknowledgements

The private enterprise cases 'The successful hawker, Dominica', 'Soex horticultural export enterprise, Senegal', 'Kallu, vegetable producer/ wholesaler', 'Enebor rice miller, Nigeria', 'Al Haji marketing board buying agent, Cameroun', 'Wimalajewa wholesale market commission agent, Colombo', 'Mafanadala produce buyer and general retailer, Zaire', and 'Matlhaku, butcher, Botswana' were taken from *The private marketing entrepreneur and rural development*. This was compiled and edited by M. Harper, Cranfield School of Management, England and R. Kavura, Institute of Development Management, Tanzania. It was issued as an offset publication of F.A.O., Rome in 1982. The 'market kings' of Sudan was written by M. Simpson, School of Economic Studies, University of Leeds. She acknowledges specifically the use of information presented by O. El. F. Addel Aziz and A. H. Khalifa in Ph.D. thesis dissertations for the University of Leeds.

'Cadbury in India' is based on information provided by A. B. Harvie Clark, P. Williamson and G. A. R. Wood. Special appreciation is expressed of the goodwill of Sir Adrian Cadbury who made this possible. 'Pepper for Tabasco' was written by G. A. Truitt, 'Trans-nationals in the development of cassava exports from Thailand' is based on a paper by Boonjit Titanivatanikan, Faculty of Economics and Business Administration. Kasetsaart University, Bangkok and on information kindly provided by A. Basler Institut für Land-wirtschaftliche Marktforschung, Braunschweig. 'Jamaica Broilers' is adapted from an article by J. Freivalds 'The growth and integration of Jamaica Broilers' which appeared in *Agribusiness worldwide* Oct./Nov. 1983. 'Unilever-Is, Turkey' was written by J. C. Abbott on the basis of information kindly provided by Mrs D. L. Wedderburn of Unilever, London and Mr M. Yildizlar, President of Unilever-Is, Istanbul. 'Mumias sugar, Kenya, Booker McConnell' is based on a case study 'Booker Agriculture International Ltd' by B. Scott, Harvard Business School kindly made available by Booker McConnell Ltd and on a paper by G. R. Allen 'The development of Mumias Sugar Corporation' published in *Oxford agrarian studies* XII(3) 1983. Appreciation is expressed of the assistance provided by W. W. Simons, Executive Director, Industry Development Council, New York, in locating and making available some of these studies and of the goodwill of the Harvard Business School in agreeing to the use of their materials in the preparation of the Tabasco and Mumias Sugar cases presented here.

'Co-operative marketing of food crops: Ogbomosho Society, Nigeria' was written by C. A. Osuntogun and I. Adeyemo, Department of Agricultural Economics, University of Ife, Nigeria. Sing Min Yeh, former chairman of the Co-operative Bank of Taiwan is the author of 'The Tsaotun farmers' association, Taiwan'. Sung Hoon Kim then Professor of Agricultural and Resource Economics, Chung Ang University, Seoul, Korea wrote 'The Ha-ee vegetable marketing group, Korea', 'Co-operative marketing of fertilizers in India: Markfed' was prepared by P. V. John, Senior Manager, Projects and Development, India Ltd, New Delhi. He acknowledges the help of R. J. Bajaj, Managing Director of Markfed, Punjab, G. S. Bhatia, Marketing Specialist of Markfed, M. S. Gulati, Sr Manager, and other officials of the organization.

'Coffee co-operatives union of West Cameroun' was written by J. C. Abbott on the basis of information collected by N. Givelet, F.A.O., Rome. 'Windward Islands Banana Growers' Association (Winban)' was written by G. Addy, O.D.A. Agricultural Economist assigned to Winban's Research and Development Division. He wishes to acknowledge the assistance of Dr J. E. Edmunds, Director of Research and Development and B. Cornibert, Economist, Winban. 'BULOG: national food grain authority of Indonesia,' was written by L. Mears, long time U.S. A.I.D. adviser in Indonesia. 'Kenya Tea Development Authority' was prepared by J. C. Abbott from the monograph by G. Lamb and L. Muller. *Control, accountability and incentives in a successful development institution* World Bank Working Paper, 1982 and the annual reports of the Authority. 'Zimbabwe Cotton Marketing Board' was written by M. J. Blackie, Professor of Agricultural Economics at the University of Zimbabwe, Harare. He wishes to thank C. G. Tracey, Chairman and R. P. N. Weller, General Manager of the Cotton Marketing Board, the Commercial Cotton Growers Association and M. Hallam, Department of Land Management of the University for the assistance they provided. H. Mettrick, I.C.R.A., Wageningen, wrote 'Botswana Meat Commission'. He was for a number of years on the board of the Commission. Appreciation is expressed of the support given by L. W. Finlay, Permanent Secretary, Ministry of Agriculture. 'Cyprus Potato Marketing Board' was written by J. C. Abbott using information furnished by P. A. Aristotelous, Chairman of the Board of Directors and A. A. Savvides, General Manager. The good offices of the Minister of Agriculture and Natural Resources, D. L. Christodoulou are hereby recognized. For the balance of the text J. C. Abbott takes full responsibility. Professor W. O. Jones, Food Research Institute, Stanford generously revised a draft of Chapter 2. Professor M. Haines, University of Wales, Aberystwyth kindly read the draft manuscript making many helpful suggestions. Special appreciation is expressed of the assistance provided by the Food and Agricultural Organization of the United Nations and its goodwill in permitting use of its materials.

Abbreviations

A.D.M.A.R.C. *Agricultural Development and Marketing Corporation, Malawi*
A.I.D. *Agency for International Development, Government of the U.S.A., Washington*
A.M.A. *Agricultural Marketing Authority, Zimbabwe*
B.A.I. *Booker Agriculture International, London*
B.M.C. *Botswana Meat Commission*
C.D.C. *Commonwealth Development Corporation, London*
c.i.f. *cost, insurance, freight, i.e. the buyer pays for a product delivered to an agreed destination*
C.M.B. *Cotton Marketing Board, Zimbabwe*
C.O.P.A.C. *Committee for the Promotion of Agricultural Co-operatives, c/o F.A.O., Rome*
E.E.C. *European Economic Community*
F.A.O. *Food and Agriculture Organization of the United Nations, Rome*
f.o.b. *Free on board, i.e. the buyer is responsible for the costs and risks of transport*
G.A.T.T. *General Agreement on Trade and Tariffs, Geneva*
I.D.A. *International Development Association, World Bank, Washington*
I.T.C. *International Trade Centre, Geneva*
K.T.D.A. *Kenya Tea Development Authority*
M.S.C. *Mumias Sugar Company*
O.D.A. *Overseas Development Administration, Government of the U.K., London*
P.F.A. *Provincial Farmers' Association*
P.F.B. *Provincial Food Bureau*
U.C.C.A.O. *Union of Arabica Coffee Co-operatives of the West, Cameroun*
Z.C.C. *Zimbabwe Cotton Corporation*

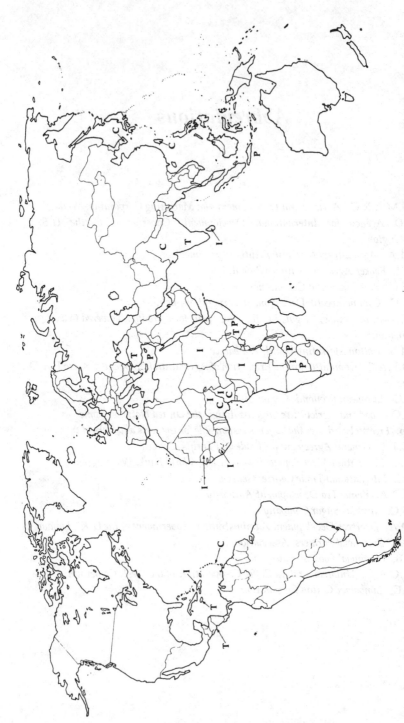

Geographical coverage of enterprises studied. Key. – I: *indigenous private* T: *transnational* C: *co-operative* P: *parastatal.*

1

Introduction

Marketing: meaning and responsibilities

In accordance with current academic practice, marketing is defined as the business activities associated with the flow of goods and services from production to consumption. The marketing of agricultural products begins on the farm, with the planning of production to meet specific demands and market prospects. Marketing is completed with the sale of the fresh or processed product to consumers, or, to manufacturers in the case of raw materials for industry. Agricultural marketing also includes the supply, to farmers, of fertilizers and other inputs for production.

Marketing tasks and responsibilities may be summarized as follows:
(a) finding a buyer and transferring ownership;
(b) assembling and storing;
(c) sorting, packing and processing;
(d) providing the finance for marketing and risk-taking;
(e) assorting and presenting to consumers.

The marketing enterprises and service agencies which are established to undertake the above functions may be independent individuals, partnerships, joint stock firms, co-operatives, mixed private and government bodies, autonomous state corporations, semi-autonomous parastatals or departments of government.

If marketing is to fulfil its role of stimulating and extending development, specific enterprises must be responsible for finding foreign or domestic buyers for the various types and qualities of produce. They must be able to arrange assembly from farms; packing and presentation in appropriate containers; sorting according to buyers' requirements; transport to buyers' depots or markets which they attend; storage to extend the availability of seasonal commodities and processing to extend the time and range of sales outlets. The enterprises must provide the necessary investment capital for fixed facilities, and the working capital to carry purchases from farmers until resale proceeds are received. Implicitly, these enterprises must possess the financial resources, the qualified managerial, sales and technical personnel, together with the initiative and willingness to accept business risks, which are necessary to perform these tasks efficiently. In export marketing, or in substitution for imports in domestic markets, they must be able to match the competence of rival enterprises in other countries.

Serving farmers, consumers and economic development

Marketing enables the agricultural producer to move from semi-subsistence to growing produce regularly for sale. Correspondingly, it allows an increasing proportion of a country's population to live in cities and buy their food nearby. Marketing also provides an incentive to farmers to grow produce for export. This increases farmers' income so that farmers form a growing market for the domestic industry. Marketing is also the means by which a country can earn the foreign exchange to pay for imports.

As consumer incomes rise, demand becomes more discriminating, i.e. a wider variety and a higher quality are sought, particularly by expanding groups of higher income consumers in urban areas. Establishment of processing industries to meet demands for new forms of products and to allow marketing over wider areas increases the complexity and scale of operations. At a later stage, competitive promotion of sales through merchandising, advertising and special services become more important.

As development proceeds, the share of economic resources devoted to the various aspects of marketing grows correspondingly in size and importance, more and more functions and services being needed to handle the agricultural produce and inputs. The channels between producers and consumers must be continually developed and broadened, or production will be constrained.

An efficient marketing sector does not merely link sellers and buyers and react to the current situation of supply and demand. It also has a dynamic role to play in stimulating output and consumption, the essentials of economic development. On the one hand, it creates and activates new demands by improving and transforming farm products and by seeking and stimulating new customers and new needs. On the other hand, it guides farmers towards new production opportunities and encourages innovation and improvement in response to demand and prices. Its dynamic functions are thus of primary importance in promoting economic activity, and for this reason an efficient marketing sector has been described as the most important multiplier of economic development (Drucker, P. J. 'Marketing and economic development' *Journal of Marketing*, Jan. 1958). The complexities of these processes and their significance for economic progress have often been under-rated, at considerable cost to economic development. National planning and investment effort has too often been focussed on production, on the assumption that, once crops are produced and roads and railways built, the development of markets and the means of serving them will be a relatively straightforward matter (Abbott, J. C. 'Marketing issues in agricultural development planning' in *Markets and Marketing in Developing Economies*, ed. R. Moyer and S. C. Hollander Irwin Inc., Homewood, Illinois 1968).

Role of government in marketing

A far-sighted government will orient its overall policy frame towards growth of those enterprises that are able to take on the necessary marketing responsibilities and will establish and maintain a favourable economic and political climate for this to happen. Major factors are: freedom to start up and operate a marketing enterprise, access to transport, banking and other commercial services, maintenance of reasonable law and order, and confidence in their continuity.

The development of a freely working marketing system can be assisted by governments through regulatory action and support services. Most governments also see some direct intervention to be in the public interest.

Regulatory and facilitating services

Marketing proceeds more smoothly and cheaply when the central local government is able to protect those involved against violence, theft and extortion. The use of weights, measures and quality descriptions understood by market participants reduces the area of disagreement and allowances that must be made for risk of fraud. Public provision for the enforcement of contracts and penalties for non-completion is a basic requirement. Fertilizers and pesticides offered for sale should show the contents on the package, with penalties for misrepresentation.

Provision and maintenance of roads, bridges, and other needs of transport, together with communication services, are services to marketing that are expected of a government. Provision of organized markets – local assembly, wholesale, retail – at convenient places is a similar central or local public responsibility. The assembly and dissemination of information on crop prospects and prices and supplies in producer, wholesale and retail markets is also recommended because it can usually be undertaken more efficiently on behalf of all market participants than by any one individual. The development of a banking system oriented to financing marketing operations may need government assistance.

Direct government intervention

Most governments intervene in the pricing of agricultural products because of the sharp fluctuations that occur under free market conditions. These are primarily due to variations in output in response to weather and other conditions that affect production. Adjusting the plantings to correct an over- or under-supply can only have an effect for the next crop season. In the meantime the price impact on producers or consumers could be severe. Governments therefore come under pressure to take measures to stabilize the prices of important crops and food items.

Their first action might be to announce official prices to farmers and/or consumers, with penalties for non-observance. Such policies may have the desired effect in the short term but they can rarely be enforced for long unless they reflect a balancing of supplies and demand. So the next step might be to establish an official supply-and-price stabilizing agency managing a buffer stock or market separation mechanism.

Existing marketing enterprises may be nationalized and state enterprises set up to gain a specific marketing advantage, e.g. to exploit a potential monopoly of sales on certain markets, or, more generally, to bring foreign exchange earning or other strategic sectors of the national economy under direct government control. In some countries ideological considerations play an important part in this. In Africa and South-East Asia there has also been a popular reaction against foreign or alien domination of marketing functions affecting large sections of the population. 'African socialism is African nationalism', said J. and R. Charbonneau (*Marchés et marchands d'Afrique noir*, Editions de la Colombe, Paris, 1961). In pursuit of such objectives many governments of developing countries have set up a wide range of state enterprises. Governments have also given special support to co-operative marketing systems because of their intrinsic appeal, and because they are a convenient instrument for the injection of finance into farming communities under rural development programmes.

Planners of agricultural development have also influenced the establishment of co-operative systems, development authorities, or other bodies directly responsive to government instructions. At first sight this simplifies the task of the planner. He could leave aside the details of working out a system of assistance, incentives and pressures needed to induce a structure of private marketing enterprises to help implement the plan. For the country, however, the risks of failure to implement the plan are magnified because so much depends on the performance of one often untried and inexperienced institution.

Commitment to long-term plans and policies in pursuit of state enterprise and co-operative systems of no technical or practical advantage appears to have sometimes over-ridden pragmatism. Also, as pointed out by R. Spinks ('Attitudes towards agricultural marketing in Asia and the Far East' *F.A.O. Monthly Bulletin of Agricultural Economics and Statistics*, 19(1) 1970 Rome), information on the performance of maketing systems that would be contrary to the official policy line has often been suppressed or never channelled on to decision-makers. In the view of experienced advisers, many governments of developing-countries have intervened arbitrarily in marketing where they should not have and have been weak where their support should have been solid and consistent.

Market conditions are continually changing: large numbers of producers and consumers are intimately involved. These two groups, and the people who earn a living from marketing, often appear to have conflicting interests. So agricultural marketing problems are frequently in the public eye. Many can be solved by spontaneous action within a flexible economic system. Some, however, persist to the point where they attract government intervention. To handle them well, a government must have its own qualified marketing information and policy analysis service at hand, and should not be unduly influenced by current political pressures.

Alternative marketing enterprises

The marketing enterprises operating in the developing countries offer a choice in the four broad categories:
1. independent, locally based private firms;

2. transnational companies with access to processing technology and external markets;
3. co-operatives or farmers' associations; and
4. marketing boards, state trading corporations and officially sponsored development agencies.

In considering the relative advantages of these categories for particular sets of conditions and requirements, specific attention must be paid to the agricultural and food marketing environment of the developing countries. Agricultural marketing is always complicated by the diverse nature and form of the products to be handled, by their perishability, by the scattered nature of agricultural production and, in most tropical countries, by the very large number of separate production units. These units can be very intensive, employing both intercropping and vertical cultivation techniques: in some ecological zones the ground can be cropped continually with plants rooting at various depths. Above ground would be shade-seeking bush crops like cocoa and above these there could be mango trees and coconut palms, with pepper vines trained up their trunks. This calls for a marketing system which is able to supply fertilizers and other inputs at the right time and one which can absorb the marketable outputs smoothly and profitably. A marketing structure able to cope with these agricultural requirements is likely to be much more complex than needed in many developed countries where agriculture is often specialized down to a very few products in any one area.

The requirements of the consumer market will also differ. Some consumers will have very low incomes and many will have exacting tastes and preferences. In highly competitive markets with quite different standards and preferences from those at home, foreign exchange must be earned. Raw materials must be provided for domestic industrial markets which have their own process-derived quality requirements and payment capacity set by the income levels of their consumers.

Governments will also be called upon to help the poorer and more socially-handicapped stratum of farmers. For them, the main considerations will be not how effective a marketing system is at the sales end but how they are treated at the local buying stage when they have produce to sell. For example, how much the first buyer does to help them match the quality and presentation of other larger farmers in the same area; how much he helps them obtain the inputs needed for efficient production and sale, and how promptly he pays and provides credit to cover cash outlays until sales proceeds come in.

For these reasons, considerable individual initiative, decision-making and skill are required in marketing. Effective structures are generally flexible in operation and allow a lot of scope for local knowledge and experience.

Indigenous private enterprise

The private marketing enterprise is one in which the capital is owned directly by the managers of the enterprise; by these managers in partnership with others, or by private investors who have acquired shares in such a company.

In most countries private enterprises are the most numerous of those engaged in marketing food and agricultural products. They range in size and sophistication from the street hawker who obtains a batch of produce on credit which he offers to passers-by until it is cleared to a woman in Ghana who meets at the port and distributes to retailers most of the fish landed, to a commission agent on the Delhi wholesale market who at any one time has some $400 000 out on credit for bananas destined to reach him by special trains.

Marketing enterprises that are operated by a single person, a family, or by partners who trust each other, can be set in motion by an informal personal or joint decision. Those involving shareholders with limited liability have to be registered under legislation governing the establishment of private companies. Generally this can be done on payment of a small fee.

Whereas a country like Nigeria is full of people intent on setting up commercial enterprises, there are many areas in the tropics where cultural traditions are against this. The energies of the socially-dominant people are directed away from marketing. This leaves the way open for outsiders to come in and acquire a strong position. Well-known examples of this are the Chinese in South-east Asia, and the Asians in East Africa. Almost inevitably this promotes resentment on the part of the majority, who feel excluded.

Transnationals

These may be defined as enterprises which produce and/or buy and sell goods in countries foreign to their own headquarters. In classical economic theory, the movement of goods from where they are plentiful and cheap to where they are scarce and expensive has been regarded as beneficial to both supplier and receiver. The movement of skills and capital to promote such trade is an extension of the same beneficial process. By the dissemination of money and skill, transnational companies can make a special contribution to the levelling up of commercial and agricultural competence between one country and another.

International private marketing enterprises have done much to foster development in the tropical countries. They are able to recruit skilled technicians and managers from the places where they are most plentiful. They can mobilize capital from sources of lowest cost. Because of their direct contact with buyers in destination markets, they are acutely conscious of competitive quality standards and techniques of presentation to consumers. It is largely due to these enterprises that export marketing of tea, coffee, sugar, bananas and rubber, for example, is relatively well organized. They have also been the introducers of large-scale retailing techniques in many developing countries. Kingsway Stores of Unilever, and the supermarkets of Printana and Monoprix have set high standards in the cities of West and Central Africa and have provided a model for local organizations to follow.

It was probably recognition of the sheer size of some of the transnational corporations that provoked a wave of public concern in the 1970s. Transnationals were nothing new. There were the British and Dutch East India companies, the Danish East Asiatic Company which *inter alia* operated the officially sponsored export produce marketing company of Liberia for a number of years. It was an awareness that the turnover of a single company such as Nestlé exceeded the gross national product of Portugal, and that financing a coup to overturn the government of a Central American republic meant little to the profit and loss account of United Fruit Co., that stirred the popular imagination.

In practice there are many constraints on the use of such financial power. First, the bulk of the transnationals engaged in food and agricultural marketing are not large, and for each of them, there is a competitor ready to step in and take its place. Nor are the transnationals necessarily a symbol of north–south dominance. There are now third world south–south transnationals also. Apart from mining and petroleum, Hong Kong is the second largest external investor in Indonesia.

Co-operatives

It is important to remember that the main aim in developing marketing co-operatives is not to pursue the 'co-operative spirit', but to create enterprises able to carry out necessary marketing functions more effectively than the available alternatives. The distinguishing feature of the co-operative is that it is owned by those who use its services and that they are entitled to share in any profits it makes. It is managed democratically by the owner-members; a committee elected on a one-member–one-vote basis directs its affairs. It can appoint a manager, but he remains responsible to the committee.

The purposes of establishing a co-operative for agricultural market-ing have been viewed from various angles. In 1908 R. C. Fay (*Coopera-tion at home and abroad*, P. S. King and Son) saw it as '. . . an association for the purposes of joint trading originating among the weak and conducted always in an unselfish spirit, on such terms that all who are prepared to assume the duties of membership may share in its rewards in proportion to the degree in which they make use of their association.' This reflects the motivation behind much spontaneous co-operative action – the desire of those who feel exploited by others to find a solution under their own control. Thirty years later H. B. Babcock framed his uncompromising definition: 'A co-operative is a legal, practical means by which a group of self-selected, selfish capitalists seek to improve their individual economic position in a competitive society'. ('Co-operatives, the pace setters in agriculture' *Journal of Farm Economics*, 1935.) This reflects fairly accurately the ideas of those groups of producers of specialized crops who, armed with a govern-ment authorized monopoly, systematically raise the price charged to hapless domestic consumers for their produce. A further 30 years later, with boom following depression, Helmberger wrote a classic article ('Future roles for agricultural co-operatives' *Journal of Farm Economics*, 1966) arguing – in the spirit of Fay – that co-operatives are formed only in conditions of revolt against the prevailing market environment. He draws the conclusion that one should expect waves of co-operatives, especially in depressed times, followed by waves of co-operative failures when the prices obtainable through other outlets improve.

While these hypotheses were generated primarily from experience in developed countries, that of the developing countries has differed only in degree. Co-operative marketing systems were established because farmers felt they were being cheated by private enterprise traders. Later, they sometimes found that the margins taken by the co-operatives, especially where assigned a legal monopoly, became too high for the service offered. 'Grey' private enterprise outlets became the more attractive.

Parastatals

These include marketing boards and their equivalents in francophone and other language countries, state trading corporations, and government-sponsored development authorities that engage in marketing. They are set up by government direction with government capital. While they may be autonomous in day-to-day operations, major operating decisions are subject to the approval of a designated government minister who will also nominate representatives to the board of directors.

In recent years all agricultural marketing bodies established by governments, perhaps with some producer and commercial representation on their directing committees, but nevertheless dependent on the government for financing and subject to its policy directives, have become known as parastatals. These are bodies for which the state is responsible, but which are autonomous in their handling of funds, staff recruitment and operational decision-making.

The objectives in establishing these enterprises are

(a) raising the bargaining power of agricultural producers in domestic or export markets via a monopoly of sales;

(b) raising the scale of operation and setting up needed marketing and processing facilities;

(c) achieving a more advantageous adjustment of the quantities and types of produce sold on particular markets;

(d) equalizing returns from sales in different markets or through different outlets; and

(e) cushioning the impact upon producers and consumers of sharply fluctuating internal and external prices.

Situations contributing to the establishment of additional parastatals in some countries, particularly in Africa, in the 1970s include:

1. conscious policy of buying out or taking over (a) foreign-owned marketing enterprise(s) such as general import/export firms and specific commodity marketing/processing enterprises,

2. concern to control foreign exchange leakages via the export marketing transactions of private enterprises; and

3. a need to set up a suitable body to handle external aid in cash or kind, for a particular area development or programme.

In the majority of developing countries the initiative to establish state marketing enterprises has been pragmatic – a response to a particular problem situation. In some countries the pattern of such enterprises has reflected a succession of influences. In Madagascar, foreign export marketing companies were taken over as national monopolies in the late 1960s. Regional bodies with authority to market rice were set up in the 1970s, and area development bodies with marketing functions were subsequently established under aid projects.

In some countries, governments or government-owned development banks, along with private investors, contribute capital to set up a marketing enterprise considered essential for development. The intention may be to withdraw the government capital when the enterprise has proved itself viable. There are also marketing boards, stabilization funds and advisory bodies which have an important influence on marketing but which do not engage in marketing on their own account. To focus on the main issues, this study will be confined to the parastatal that has special responsibilities or monopoly powers assigned by the government for buying and selling produce.

Issues for discussion

1. What does marketing involve? What is the role of the marketing enterprise?
2. How does marketing benefit producers? Consumers? Economic development?
3. How can governments foster marketing development? Illustrate this from experience in your own country.
4. What are the main factors leading to government intervention in marketing. Examine the various attitudes towards this in your own country. What trends in these attitudes do you envisage for the future?
5. What are the main forms of marketing enterprise to be found in your own country? Present profiles of examples of each type.
6. What influences have been instrumental in shaping the forms of marketing enterprise in your country? Are these likely to continue? What changes do you foresee and for what reasons?

Further reading

Abbott, J. C. and J. P. Makeham, *Agricultural economics and marketing in the tropics*, Longman, 1979.

Abbott, J. C. *et al.*, *Marketing improvement in the developing world*, F.A.O., Rome, 1984.

Abbott, J. C. and H. Creupelandt, *Agricultural marketing boards; their establishment and operation*, F.A.O., Rome, 1966.

Arhin K., P. Hesp and L. van der Loan, *Marketing boards in tropical Africa*, Kegan Paul, London, 1985.

Bateman, D. L., J. B. Edwards and C. Levay, 'Problems of defining a co-operative as an economic organisation' *Oxford agrarian studies* (8) 1979; 53–62.

Bauer, P. T., *West African Trade*, Routledge and Kegan Paul, London, 1963.

C.O.P.A.C., *Commodity marketing through co-operatives; some experiences from Africa and Asia and some lessons for the future*. C.O.P.A.C. Secretariat, F.A.O., Rome, 1984.

F.A.O., *Agricultural price policies*, Rome, 1985.

Jones, W. O., *Marketing staple food crops in tropical Africa*, Cornell University Press, Ithaca, 1972.

Livingstone, I. and H. W. Ord, *Agricultural economics for tropical Africa*, Heinemann, New York, 1981.

United Nations Centre on Transnational Corporations, *Transnational corporations in food and beverage processing*, U.N., New York, 1981.

2

The marketing enterprise and economic development

Any presentation of principles and models for the behaviour of marketing enterprises and their contribution to the overall development of an economy has to begin with the theory of pricing. This provides a basis for predicting how an enterprise will respond to a particular marketing situation and the direction into which pursuit of its best interests will lead it.

The determination of market price

In the market, demand and supply come together. The market provides a place where those who wish to sell quantities of meat, or grain, or whatever, can expect to meet people who would like to buy. Because individuals' wants for money and for various commodities differ, amounts for sale and the prices that people will pay for them will differ. What a particular individual will buy or sell will vary with the price.

The total amount that participants in a market are willing to sell at various prices may be represented graphically by a 'supply curve' and the total amount that participants are willing to buy by a 'demand curve' (Fig. 2.1). Where these two curves intersect, the amount that is offered for sale and the amount that people are willing to buy is the same. This is the equilibrium price or market-clearing price. At a higher price, more goods will be offered for sale than will be purchased; at a lower price some buyers will be unable to buy what they want. At the equilibrium price the market clears: at this price all who wish to sell can find buyers and all who wish to buy can find sellers.

In any market where prices are freely determined, they will move toward equilibrium. At a price slightly higher, some goods will not be sold and some owners will offer to sell at a lower price. At a price slightly lower, some buyers will not be satisfied and will offer a slightly higher price. The balancing of supply and demand at a market-clearing price may be illustrated with the following example for meat:

Price per kg (S)	Meat demanded (kg)	Meat supplied (kg)
2.50	85	200
2.00	95	185
1.50	120	120
1.00	140	60

A price of $2.50 per kg will attract plenty of meat (200 kg) but consumers will only buy 85 kg – the price will rise as consumers bid against each other for some of the 85 kg that are offered. As it rises, more suppliers will be willing to sell, until at $1.50 just as much meat is offered as consumers are willing to buy. Similarly, if prices differ in neighbouring markets, merchants will buy in the market where prices are lower and sell in the market where prices are higher. This process of buying cheaply in one market and selling expensively in another until prices between the two differ by no more than the cost of transport is known as arbitrage. Arbitrage also occurs over a time period, by storage, and between different forms of a commodity, by processing.

Fig. 2.1. Balancing supply and demand at an equilibrium price.

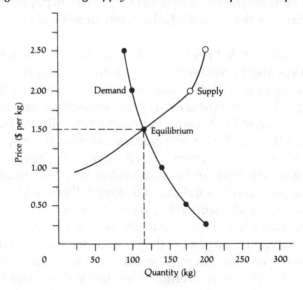

Competition

In a fully competitive situation a firm can sell all its goods at the market price, but it cannot, by its actions, affect that price. If it asks a higher price it will sell nothing. If it offers its goods at a lower price it will receive more orders than it can handle. It will attract orders from all other sellers and force them to reduce their own prices. The result of this will be that equilibrium is re-established. The extent to which the quantity of a product supplied or demanded is affected by changes in price is known as elasticity. Under full competition, the demand curve faced by an individual producer or seller is infinitely elastic (see Fig. 2.2). This is because of the small share he contributes to the total market. However, the demand faced by suppliers as a whole will also be related to price. The shape of the supply curve is likely to be nearly the same for all suppliers as it is for individual suppliers. Cost of production does not have much short-term influence on price. This highly inelastic supply condition is responsible for the rapid falls in price that sometimes occur in free agricultural markets.

In principle, competition encourages efficiency. Under competition, all a firm can do is try to cut its costs to a minimum. New enterprises can open up at any time; so an enterprise must always keep up to date with technical developments that may enable it to cut costs. Moreover, if an enterprise is able to reduce the cost of its product, it must reduce its price to the consumer. If it does not it will lose business to firms who reduce their price when they make use of the new development. This is the situation in which many small farmers and marketing enterprises find themselves.

Fig. 2.2. Individual supply-and-demand curves.

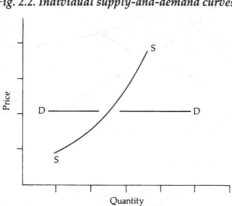

Monopoly

This is the opposite of competition and means that there is only one seller. This seller controls all the sales of a product and, through this control, can influence the price.

The monopolist will take into account the effect of his volume of sales upon price. He can choose the level of sales and price that offer the greatest returns on his operating costs. The price and output he decides on are governed by the elasticity of demand for the product.

The monopolist obtains maximum profits when the growth of output, or sales, adds just as much to revenue as it does to costs. Figure 2.3 illustrates this by showing how maximum profits are obtained when marginal revenue equals marginal cost. Marginal cost equals marginal revenue at the level of output Q1. The price is obtained from the average revenue for this quantity, i.e. P1. Firms in competition would have had to sell at the price where average cost and average revenue meet, i.e. P2. This would mean that the quantity sold, Q2 would be greater and the price obtained much lower.

Fig. 2.3. Monopolistic pricing.

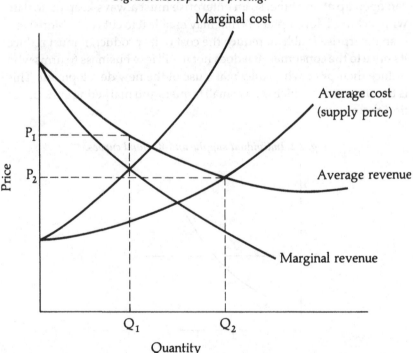

Monopsony

The opposite face of monopoly, where there is only one buyer is called monopsony. Farmers who grow produce for processing often find that there is only one processing plant in their area. This plant can, within limits, set its own buying price because it has no competitors. Under normal circumstances the more it buys the higher the supply price. Therefore the plant will buy that quantity at which its marginal cost equals the demand price. This is shown in Fig. 2.4. The quantity purchased at the price P1 is Q1. Thus it buys less and pays less than if it faced competition. Several firms competing against each other would buy the larger quantity Q2 at the higher price P2.

The limit to which a monopsonistic enterprise can keep down the prices it pays depends on several factors. One is the extent to which it can discriminate among suppliers, i.e. pay lower prices to some than others, reflecting their costs and bargaining positions. At the same time it must safeguard its own invested capital and profit opportunities by keeping its suppliers in business. The ease with which a new enterprise can set up in opposition will also be a consideration.

In various countries, abattoirs, marketing boards, state companies and co-operatives have been made monopolies by government action. Other private and public enterprises are almost monopolies because they are protected from competition by high duties, exchange control and other limitations on imports. Monopolies and monopsonies are not necessarily bad. They may be able to operate at lower costs and

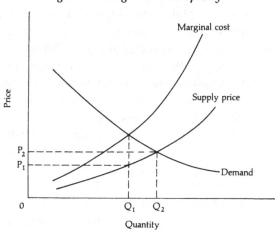

Fig. 2.4. Pricing under monopsony.

promote sales more effectively than smaller firms because of economies of scale, i.e. they may be able to employ better qualified staff, acquire more efficient equipment and undertake more elaborate sales promotion because the cost of such outlays is spread over a larger volume of business. In practice, these potential savings in cost because of a larger scale operation may be negated by diseconomies in management, i.e. by an inability to control a wide ranging set of operations effectively. Official monopolies in tropical countries are often criticized for the low quality of service and the fact that illicit payments to their staff have to be made to get service at all. Pressure on a monopoly for more efficient performance can come from influential enterprises with which it has commercial dealings, from public opinion and from the government.

Oligopoly

There is an intermediate stage between monopoly and competition in areas of marketing and farm supply which are not subject to official monopolies. This stage is called monopolistic competition or oligopoly (Greek, meaning 'sale by a few'). In many countries a few firms dominate the supply of such products as fertilizers. These firms are all aware of each others' positions and possible reactions. They tend to avoid direct competition because it would result in the competition being returned. They compete in other ways, for example, by offering credit, by entertaining their customers, by advertising, and making their products and services look more attractive.

Separation of markets

Organizations able to control the total output of a product for sale can increase their returns in a number of ways impossible for competing enterprises. They can limit the total quantity of a product put on to the market. By planned storage they can control the release of supplies. They can control the amounts sold in separate markets. And, if they can prevent movements between these markets, they can charge exactly what each independent market will bear.

Discriminatory pricing, as this system is called, takes advantage of differences in demand in different markets. The monopoly seller charges a high price in a high-income or inelastic demand market, and a lower price in a low-income market where demand is more elastic and where many people would stop buying if the price were high. These markets must be separate, i.e. geographically apart or administratively separate, otherwise the monopoly seller would have to charge the same price to all customers. This means that high-demand buyers might buy at a lower price than they would have been willing to pay; while, at the lower demand end, sales would decrease because the price would be too high for some customers.

The demand curves with differing slopes in Fig. 2.5 refer to two markets that can be separated. Both demand curves are above unitary elasticity (the section marked A). Profit maximizing prices for the seller are those where MR is equal in the two markets. Normally sales would not be expanded beyond the point where MR becomes negative; total revenue could be increased by selling less. However, an enterprise selling produce already in hand might sell in the area B where MR is negative. The most profitable prices would then be where MR is equal in the two markets, i.e. P_1 and P_2.

Separation of markets may be complete where movement of a product from one market to another is directly controlled by a monopoly marketing board specifically for this purpose, or indirectly by bodies applying disease transmission controls, as on livestock and fruit. Separation may also be complete within certain price ranges if there are tariff barriers. It may be effective, to varying degrees, where it is based on difference in appearance, grading, packaging and advertising of the same basic product. Separation of markets by this last method is possible provided that sellers can convince their customers

Fig. 2.5. Price discrimination.

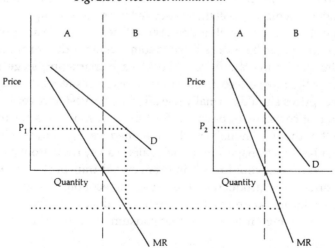

that different brands of the same product merit the payment of different prices. Individual farmers or small traders rarely sell enough to be able to use such methods. However, they may be able to build up 'sheltered markets' of regular customers who believe that their produce is especially fresh or of very high quality.

Optimum allocation of resources

The function of a pricing system is to direct resources to where they are wanted. If the price of a product is high this attracts land, capital, labour and entrepreneurship into increased production. In this way demands reflected by higher prices are satisfied. The continual adjustment of resources in response to demand is a sign of a healthy and flexible economy. It means that demands are met as they occur. It also means that resources are not used to produce foods that are not wanted – very wasteful for an economy.

The following is required for prices to guide agriculture towards producing what is wanted.

(a) The broad level and pattern of prices should be known early on so that production programmes can be adjusted.
(b) Price expectations should allow enough time for production plans to be completed with reasonable certainty.
(c) Information on prices should be clear enough to allow farmers to interpret them for their own individual situations.

The price level should encourage the amount of production that can be handled by the market at an acceptable price. If production is maintained at a higher level than is needed, by means of support prices for example, resources will not be used efficiently. The extra production capacity would be better shifted to some other product or activity.

At the same time the prices paid at each marketing stage must encourage the most effective allocation of resources to marketing. For example, prices should normally rise after the harvest to cover storage and interest costs. Also, payment for transport must be attractive to ensure that there is enough, well-maintained transport available. On the other hand, a transport monopoly, along a key route from producing area to market could result in exaggerated transport costs which would discourage production and cause the economy to suffer. The function of the marketing system is to shift produce from place to place, store it and change it to a form that consumers want. Market prices indicate what people want and thereby encourage efficient use of resources. Resources are not likely to be well allocated if the natural movement of prices is hindered by movement controls, tariffs, monopolies and arbitrary fixing of prices and margins.

Role of marketing in the development process

Marketing is necessary for economic development. Marketing facilities allow concentration on the easiest and most rewarding lines of endeavour. The products of specialization in one area may then be exchanged for those obtained more easily elsewhere. Everyone benefits when the producers of different products are able to sell their goods to one another. They can obtain, with part of their own output, more of other goods than they could have produced themselves with the same effort. Improvements in marketing procedure or organization expand trade, raise the level of living of all concerned and add to the economic wealth of the community. These were the fundamental principles for Adam Smith, the great protagonist of free trade, in the 1770s, and are still the basis of economic thought today.

Vent for surplus

Access to a market outside a particular area can be the way to make use of under used resources. Until farmers in Eastern and Western Africa, for example, were given the chance to grow export crops (through the construction of railways to the coast and the opening up of overseas markets), they concentrated on subsistence food production and traditional activities that did not fully employ available land and labour. The Mombasa to Kisumu railway, opened in 1901, reduced the cost of transporting goods to the coast from as much as $400 per ton for porterage to $3.50 per ton. Access to export markets induced peasant farmers to increase export crop production very rapidly. Incentives were heightened by the possibility of purchasing consumer goods supplied to local traders by European shippers. Defined regions within a country can become a market for food supplies when easy transport is available. Thus the forest belt and coastal areas of West Africa have been outlets for the grains, pulses and cattle produced under more favourable ecological conditions to the north.

Transnationals based in the importing country were generally responsible for organizing market movements through such vents for surplus but indigenous enterprise has also had an important role as evidenced by the Hausa trade in kola nuts.

Extraction of funds for development

A vent for surplus allows the creation of capital for investment outside agriculture. This is the basis of the extraction of the agricultural surplus thesis so beloved of the last generation of development economists. The concept is that, when there is no mineral wealth, development capital must be squeezed out of the only basic resource, agriculture. This can be done through direct taxation (as feudal dues). It is much easier when there is a profitable flow of produce to external markets that can be taxed at the point of export. In Sudan such funds were accumulated and invested by the great trading families. Many other countries used monopoly export marketing enterprises to capture 30 to 50 per cent of the export price for the government, sometimes stifling the goose that laid the golden eggs. Killick's account of government investment in Ghana during the Nkrumah period *(Development economics in action*, St Martin's Press, New York, 1978) demonstrates how wasteful such investments can be when political aspirations become the basis of decisions.

Marketing for levels of development

A number of researchers have analysed marketing systems in relation to broad levels of development. W. O. Jones ('Measuring the effectiveness of agricultural marketing in contributing to economic development: some African examples'; *Food Research Institute Studies* 9 (3) 1970) has set out stages, from subsistence farmer in New Guinea cut off from all outside markets and supplies to the agriculture of the Salinas Valley in California where all output is for the market and all inputs including land on a seasonal basis can be acquired from it. An F.A.O. study tried a statistical correlation approach. It was felt, however, that the measurable indicators of marketing development – number of transport vehicles, of telephones, etc. were not the most influential factors, apart from obvious cross-correlations. J. T. Mentzer and A. C. Samli, *(Columbia Journal of World Business,* Fall 1981) take, as the key to economic development, the construction of a production/ marketing infrastructure with productive, distributive, communicative and persuasive features. They then go on to suggest a weighting system for the contributory factors: management, transport, storage, standardization, marketing knowledge, amenability of decision-makers to marketing considerations, and so on.

The marketing enterprise and development

We have already mentioned the fact that economists have neglected marketing as a causal factor in development. They knew about trade, of course, and the importance of foreign exchange. Who generated and organized the trade and the whole process of domestic marketing was a blank on their thought maps. R. Holton's 'Marketing structures and economic development' *(Quarterly Journal of Economics* 67, August 1953) and, with J. K. Galbraith, *Marketing efficiency in Puerto Rico* (Harvard University Press, 1955), were among the earliest studies to focus on marketing *per se* as a causal factor in development. The first development economist to recognize this explicitly was W. Rostow (*'Marketing and economic development'*, P. D. Bennett ed., American Marketing Association), who concluded that the most important factor for further economic growth in an underdeveloped national economy was the development of a national market system. The first marketing scholar to see marketing as the driving force in economic development was Peter Drucker, who takes a managerial approach to development. Marketing promotes development by creating standards, and by developing managers and entrepreneurs ('Marketing and economic development', *Journal of Marketing*, January 1958). In the same issue of this journal Bartels showed how marketing stimulates economic development, by changing behaviour, attitudes, relationships and values, instead of by just adapting to it.

Charles Slater is now seen as the prophet of the marketing enterprise being a catalyst in development. (Nason, R. W. and P. O. White, 'The visions of Charles C. Slater', *Journal of Macro-marketing* (1) 1981). He put forward three hypotheses.

1. Horizontal market co-ordination by multi-product retail marketing organizations can cause real income to be redistributed by offering widely consumed necessities at lower prices than goods consumed mostly by higher-income households, so stimulating output and improving welfare.

2. Vertical market co-ordination by marketing enterprises can reduce sales risks and so encourage suppliers to increase output and so bring down their costs.

3. 'Social' marketing by marketing planners at the macro-level involving the use of existing marketing institutions and channels can achieve changes in attitude and behaviour of segments of society and likewise improve welfare.

Hypothesis 1 focusses on the creation of multiproduct outlets with high turnover and low margins. Hypothesis 2 hinges on vertical marketing co-ordination bringing lower prices and increased efficiency. The result would be lowering of prices in urban areas, which would encourage more farm and convenience-good production. Both of these hypotheses require action at the micro-level by individual entrepreneurs. One of the most telling bases of this thesis was a case where a food retailing chain in Recife, Brazil, acquired a rice mill in a distant producing area. Because it could absorb the supplies coming from this mill and also count on them and plan their flow directly, it was able to reduce its retail price in Recife by $5.0 per ton. Its market share expanded sharply as a result so that eventually competitors had to bring down their prices to around the same level.

Depending on the relative bargaining power of the parties concerned, the benefits of such a saving due to better marketing organization are as follows:

– consumers are able to purchase more rice with the same income, to spend more on other consumer goods, or to accumulate savings for investment;
– producers have the incentive to expand production, improve incomes and levels of living and savings for investment;
– the marketing enterprise makes increased profits until they are eroded by competition, has the incentive to expand operations, and capacity to make further investments in this or other lines of business.

That micro-marketing enterprises have made, and continue to make, such contributions to development is clear enough. In his 'A macro marketing approach to economic development', (*Journal of Macromarketing* (14) Spring 1982), E. W. Cundiff goes on to ask how the macro-marketing planner can achieve such micro-level changes to order. The very act of the central planner in initiating change may reduce the initiative of the individual entrepreneur. He may come to expect the central government to solve his problems rather than finding imaginative solutions of his own. The macro-marketing planner must find ways of developing a climate that encourages, and eases, the way for individual initiative.

Cundiff's own public policy agenda for micro-level development initiatives is the now conventional western marginalist approach.

1. Change prevailingly negative attitudes towards marketing and participation in it.
2. Develop marketing curricula in national educational institutions.
3. Promote specialized practical training of marketing managers.
4. Invest in marketing infrastructure and support facilities – transport, storage, communications, credit and information.
5. Strengthen the range of assembly, wholesaling, retailing and integrated enterprises by maintaining freedom of entry and easy access to finance, together with measures to restrict collusion.

Directly counter to this is the Marxist view.

The Marxist view

Interpreters of what Marx wrote are legion. In regard to marketing, his followers have gone along with the common man's short-sighted view that there is no 'product' from the 'middleman' between producer and consumer. So there is no place in the fully Marxist economy for the enterprise, or the man, who buys from producers where their product is plentiful and cheap and sells it to consumers where it is scarce and dear. The value accruing to society from a better balancing of demand and supply according to consumers' preferences has low priority. Since there must be a movement of goods from producers to consumers, this should be undertaken by agencies of the people – central or local government enterprises or co-operatives that will not be doing it for a profit. Prices are set by central decisions taking very little account of quality differences, seasonal scarcities, etc.

Because of practicalities, this system cannot be extended to every detailed transaction. Agricultural produce in China, for example, was classified – until the relaxations under Deng in 1981 – into three categories, according to their importance to the national economy and people's livelihood.

Category I (state monopoly): grain, edible oil, timber and cotton were procured and distributed directly by the government. As an incentive to production, quantities offered by farmers after fulfilling a set quota were bought by the government at a higher 'above-quota price'.

Category II (purchase by contract): hemp, silk, tobacco, tea, sugar, fruits and vegetables, fishery products, meat, poultry, eggs and medicinal herbs, up to a specified quota had to be sold to the government in accordance with the state plan. The balance could be sold to the state or to other markets as producers wished. Most cities had special municipal companies to wholesale vegetables, fruit and poultry which made contracts with producer groups.

Category III (open sale): other products might be bought by the Ministry of Commerce by negotiation; however, producers could sell these on free markets.

There was advertising in China in the late 1950s and early 1960s. It was condemned as a symbol of capitalism by the Cultural Revolution in 1967 but began again with the economic reform of 1979.

Congruence of Marxism and the development planners

Limitations seen in current prices as signals for longer-term investment, especially in developing countries, led in the 1950s to a consensus in favour of development planning. In the words of Scitovsky, 'Two concepts of external economies', *(Journal of Political Economy,* April 1954) 'the proper co-ordination of investment decisions requires a signalling device to transmit information about present plans and future considerations as they are determined by present plans; the pricing system fails to provide this. Hence the belief that there is need either for centralized investment planning or for some additional communication system to supplement the pricing system as a signalling device'. In 1957 Gunnar Myrdal wrote that it was 'universally urged' that low-income countries 'should have an overall, integrated national plan'. 'Because of the various deficiencies in a backward country it is also accepted by everyone that the government will have to take over many functions which in most advanced countries in the Western world were left to private business.' *(Economic theory and under-developed regions* Methuen, London, 1957).

There was a dissident school with P. T. Bauer its most vigorous spokesman. His attacks on such hallowed notions as the vicious circle of poverty, central planning and international aid, and his defence of the decentralized decision-making process of the market were at first too strongly expressed to command much support. By the late 1970s, however, they were in fashion along with the Chicago school of Friedman. Experience of centralized planning in practice had led to disillusionment. The information on which it was based was inadequate and its administration inefficient. The distorting effects of import and export controls and over-valued exchange rates became evident. Interest shifted back to the efficient allocation of resources and to comparative advantage.

Bread and palaces

The political leaders of many newly independent countries found state enterprises very convenient. They can be made physically very impressive; they give the impression that the government is doing something; they can be spread out over various parts of a country, create new jobs there and provide excellent opportunities for patronage.

The fragility of the governments of many developing countries in the face of temporary economic crises; the knowledge that an increase in the price of foodgrains to urban consumers could mean riots in the streets; the awareness of their susceptibility to an internal coup by a group of army officers with summary penalties for 'hoarders' and 'profiteers' to follow, and the continuing preference of the ruling group for import controls that opened the way to direct financial rewards for import licences and other patronage all tend to impede the working of the marketing enterprise as a mechanism for development. An enterprise cannot be allowed to follow its own course and achieve its full potential because growth, independent of the government, is seen as a threat to the government.

These various aspects and consequences of weakness in government show up in:
1. maintenance of government export monopolies to collect discriminatory taxes on export agriculture;
2. preoccupation with food security and attendant distortions in foodgrain marketing systems; and
3. import control regimes based on arbitrary allocation of licences to favoured individuals.

All of these have the effect of favouring narrow interest groups. They work against the development of marketing enterprises capable of operating on a national level. Monopoly profits are still there, but they do not go to an enterprise likely to invest them in achieving further economies of scale and productive capacity; they go instead into patronage, consumption of luxury imports, and into foreign bank accounts.

Multiplier effect of the marketing enterprise

How much impact on the pace of development does the marketing enterprise have? Lack of an accepted indicator for this may be one of the reasons why the marketing enterprise is neglected in development economics.

This concept has to be seen in terms of different levels of development of a country and economic orientations, and in terms of alternative marketing structures. Thus the impact that the efficiency of the marketing enterprise will have on economic development will have to be evaluated separately for countries with:

(a) the bulk of the population still resident in rural areas;
(b) many people shifting to urban areas and becoming dependent on purchased food supplies;
(c) continuing dependence for foreign exchange earnings on the export of agricultural products;
(d) fairly well-balanced, but stagnant, economies needing new points of incentive to stimulate growth.

Certainly some of the risks can be assessed. The marketing of smallholder crops in Malawi is dominated by the parastatal ADMARC. It has been reasonably well run. This has been a positive factor in the country's development. If it were not efficient the negative impact would be great indeed, with the fall of the government a probable consequence. Has its predominance, however, stunted the growth of indigenous private enterprise which would have offered a wider range of local initiatives? In Nigeria, urban concentrations of several million people were supplied effectively during successive stages of growth by an array of indigenous marketing enterprises.

Examination in depth of the contribution to development of alternative marketing enterprises under differing conditions is a research area for the future.

Issues for discussion

1. How are prices determined? What is meant by an equilibrium or market price?
2. What is the nature of the competition in the marketing of grains, other crops, livestock, fertilizers in an area you know? Have there been significant changes in the share of the total market taken by different enterprises? What were the reasons for these changes?

3. Do you know of any marketing enterprise that has a monopoly at the village level, the assembly market level, the national level? On what is this monopoly founded and what are its limits?

4. Assess, for some of these enterprises, the advantages that they obtain from their monopoly. What are the advantages and disadvantages for the farmers concerned? The consumers? The economy of the country?

5. Illustrate, from experience with some products of your country, the principle of market separation. What are the limits of this separation? What advantages accrue to growers? To the economy of the country?

6. What issues are at stake regarding optimum allocation of resources when there is competitive pricing? Pricing by a monopolistic private enterprise? Pricing by government decision?

7. What do you understand by the vent for surplus concept? How has it applied to your country? To whose initiative do you attribute it? How were the benefits distributed?

8. What have been some of the main contributions to thought on the role of marketing in development? On the marketing enterprise in development? Illustrate these ideas from experience in your country.

9. Assess the implications for a country you know of the Marxist view of marketing. How has implementation been adapted to the realities of practical marketing operations? What have been the advantages and disadvantages, in particular product areas, to farmers? Consumers? And the economy?

10. Set out ways in which a marketing enterprise can have an accelerating effect on the pace of development. What would cause it to have a braking effect and how would this be manifested? Illustrate, if possible, with enterprises you know.

Further reading

Abbott, J. C. 'Technical assistance in marketing: a view over time', *Working Papers: Marketing and rural development*, German Foundation for Economic development, Feldafing, 1977.

Abbott, J. C. and J. P. Makeham, *Agricultural economics and marketing in the tropics*, Longman, 1979.

Bateman, D. L., J. R. Edwards and C. Levay, 'Agricultural co-operatives and the theory of the firm', *Oxford Agrarian Studies* (8) 1979: 63–81.

Bates, R. H., *Markets and states in tropical Africa: the political basis of agricultural policies*, Berkeley, University of California Press, 1981.

Bauer, P. T., 'The economics of marketing reform', *Journal of Political Economy* 62(3), Chicago, 1954.

Bauer, P. T., *Equality, the third world and economic delusion*, Weidenfeld and Nicolson, 1981.

Harper, M. and R. Kavura, *The private marketing entrepreneur and rural development*, F.A.O., Rome, 1982.

Kindra, G. S. ed., *Marketing in developing countries*, Croom Helm, London, 1984.

Samuelson, P. A., *Economics: an introductory analysis*, McGraw Hill, New York 1973.

Slater, C. C., 'Market channel co-ordination and economic development' in *Vertical marketing systems*, ed., L. P. Bucklin, Scott, Foreman & Co., Glenview, Illinois, 1970.

Stigler, C. J., *The theory of price*, Macmillan, New York, 1969.

Timmer C. P., W. P. Falcon and S. R. Pearson, *Food policy analysis* Johns Hopkins University Press, 1983.

United Nations Department of Economic and Social Affairs, *The multinational corporations in world development*, U.N., New York, 1974.

Widstrand, C. G., ed., *Cooperatives and rural development in East Africa*, African Publishing Co.. 1970.

3

Indigenous private enterprise

The successful hawker – Dominica

Farming is the chief activity in Dominica, with only a few people employed in other activities like construction, transport and trading. The main crops are bananas, limes, coconuts and grapefruits; other citrus fruits, cocoa, mangoes, avocados and potatoes are also grown. Some of these crops are processed on the island to make lime juice, lime oil, copra, rum, cooking oil and soap. A quarter of the island's 80 000 people are unemployed.

Those who are lucky enough to acquire technical skills obtain the few jobs available. The remainder have to fight for survival. Those who have some entrepreneurial inclination very often become 'hawkers', i.e. traders in island produce.

Mary Jane left school in 1977 and tried desperately to find a job. She was willing to teach, to be a secretary, a shop assistant or a messenger but there were no openings. In desperation Mary Jane thought of going into farming. She looked into the market for agricultural produce before starting an enterprise of her own; she soon discovered that there was a surplus of food. This led her to look at opportunities for marketing this surplus as a hawker. There was a government agricultural marketing board, but it was not able to dispose of all the fruit, vegetables and flowers for sale. Mary Jane believed that there was a market for this produce in other islands, and that it would be possible to operate year-round business without too many seasonal fluctuations.

Encouraged by what she had learnt, Mary Jane decided to go into business. She would buy surplus produce in Dominica, take it to Guadeloupe, or more distant islands, and sell it there. First, she had to estimate how much to take, and how much money she would need to pay for it and cover the transport and other expenses. She found that she would need $250. She invested $125 of her own and borrowed the balance from the revolving fund of the Youth Development Division of the Ministry of Education and Health.

Mary Jane was meticulous. She decided that everything she sold would be top quality. Her own business would depend on the appearance of what she had to sell and her produce would, in a sense, be an ambassador for her country. She bought her produce, supervising the entire process of harvesting, packing, storage and preparation for shipment. In September 1978 she flew to Guadeloupe on her first trip as a hawker. It was very successful; she sold her produce for $650 and was left with a profit of $400. The produce had cost her $150, transport for herself and the produce cost $40. The balance was spent on accommodation in Guadeloupe. Naturally, she was very pleased with these results and made two more trips. She then started buying things in Guadeloupe which were needed in Dominica and was thus able to earn money in both directions. In time she employed somebody in Dominica on a part-time basis to find people who wanted goods from Guadeloupe, and also to supervise the collection of produce while she herself was trading in Guadeloupe.

After the third trip she paid off the original loan of $125, replaced the similar sum she had drawn from her own account and was left with

Table 3.1. *Income and expenditure accounts; Mary Jane, Dominica.*

	$	$
Income		
Sales from hawkering		7000
Sales in Dominica		2500
		9500
Expenditure		
Purchase of produce	2000	
Packing and agent's fees	400	
Purchase of goods in Guadeloupe	1625	
Shipping and related charges	150	
Board and lodgings in Guadeloupe	750	
Wages and related overheads	900	
Air passages	300	
Total		6125
Net profit		3375

$1400 as operating capital for further trips. Mary Jane's results for 1979 are summarized in Table 3.1. At this point she had $3750 in cash and a further $200 worth of items bought in Guadeloupe and not yet delivered to customers in Dominica.

Mary Jane continued her successful operations until the hurricane in August 1979. This destroyed all the agricultural produce of the island and put an end to her business venture. Mary Jane had by then, however, made a number of contacts in Guadeloupe: she was able to obtain employment in a key position in a business to which she had previously sold produce. She became a part-owner of this business and looks forward in the future to resuming trade with Dominica when conditions are favourable.

Not unreasonably Mary Jane is pleased with her venture; she is earning substantially more than would be possible in the kind of employment she would have been able to obtain, had any been available. She has created one full-time job in addition to her own, and she has provided part-time employment for a further three people. In addition she has developed a market for a large number of farmers and has helped many people in Dominica to obtain things from abroad which they could not afford to go and buy personally.

Soex, horticultural export enterprise – Senegal

Market gardening in Senegal started in the 1950s. It was small in scale serving the residents of Dakar, the capital city. The next step was to produce for the winter markets of Western Europe. In 1972, the vegetable marketing subsidiary of the transnational Castle & Cooke Inc. and the Government of Senegal established B.U.D. Senegal. It planned to occupy over 3600 hectares and to export 80 000 to 100 000 tons of produce annually. For a number of reasons B.U.D. Senegal never realized its expectations and was closed. There are now about 15 much smaller vegetable export traders: the largest export 1000 tons of vegetables per year. Soex is one of these export traders. The enterprise was started in 1970 by Abdoulaye Diallo. It was set up as a limited company with an initial capital of $52 000. By 1980 this had increased to $167 000. During its first season Soex bought 100 tons of vegetables from 12 groups totalling 200 or 300 peasant producers. By 1980 it dealt with 22 groups representing between 600 and 700 peasants and exported over 600 tons. Most of this was green beans produced between December and June.

Organization

Soex is represented in the villages by a Sector Chief. He makes contacts with farmers on behalf of Soex during the month of September and distributes seeds to be planted during the first or second week of October. Seeds are sown every 20 days until the middle of March. The first beans are harvested at the beginning of December and harvesting continues until the end of May. They are picked every day. To collect them Soex has seven lorries each of between 5 to 7 tons capacity. One of these is refrigerated.

The farms deliver their produce straight from the field. To meet the standards of European customers, broken, blemished and insect-damaged pods are eliminated. The beans are then graded into the finest grade (less than 6 mm), fine grade (6–9 mm) and those above 9 mm which are of normal grade. Usually only the finest and fine grade fetch a price sufficient to cover the cost of packing and air freight.

All this involves a great deal of work. There are 18 full-time employees and 200 seasonal labourers employed for storing, sorting, grading and packaging: they are organized into two teams in order to ensure maximum productivity. The equipment used includes ten sorting tables, three bean grading machines, 60 packing points, a multi-stage weighing machine and two cold rooms each of about 100 cubic metres. Soex purchases about 200 000 cartons every season.

Marketing

The day's harvest is dispatched the same night or the following morning to Paris, Marseille, Lyon, Nice and Bordeaux, Brussels, Vienna and Geneva. Soex has representatives in each of these markets, usually wholesale commission houses. Generally they take a 10 per cent commission on their sales.

A range of prices to producers is agreed in meetings between the traders' association, farmers' representatives and government officials. In 1980 this was 48 to 67 cents per kilo according to market conditions. Typically, Soex paid about 57 cents if grading losses did not exceed 20 per cent. Down-graded beans cannot be sold at all, so Soex effectively pays about 70 cents a kilo for export quality beans. Itemized expenses per kilo are set out in Table 3.2.

The air freight rate in 1980 was 74 cents a kilo to Paris, $1.0 to other destinations. Packaging increases the weight by about 10 per cent so the total freight charge to Paris was 81 cents a kilo. Customs charges and transport from the airport to the wholesale market totals around 10 cents per kilo making a total of $2.10 per kilo. Considering the selling commission of 10 per cent and the cost of storage, accidents and problems during loading and unloading, it is clear that any sale for less than around $2.50 a kilo will involve a loss. Monsieur Diallo recognizes that he depends on the small farmers who supply the vegetables, so his policy is to pay prices sufficient for them to make a reasonable living even when the market is low. He hopes that he can cover the losses that this involves by making a more generous profit when sales conditions are favourable.

The commission selling system means that Soex has to allow its customers credit: it pays its suppliers regularly every week. Access to substantial working capital is essential, particularly during the beginning of the season when vegetables are being bought but no cash has come in from the market. Soex uses the funds at its own disposal and supplements these with credit from the bank. M. Diallo keeps the bank loans as low as possible because the interest charges are high. Although he wants to increase the business he is limited by his financial capacity: he has to balance his own ambitions for growth with his limited financial resources.

This type of business requires rapid decision-making and speedy implementation this imposes unusual hours of work and methods of operation. During the production season M. Diallo works more than 16 hours a day. Flights leave every day including Saturday and Sunday. Fresh vegetables have to be dealt with immediately: anything kept in Senegal for more than 24 hours loses much of its value.

Table 3.2. *Supply, packing and transport costs of green beans; Soex, Senegal.*

	$ per kg
Effective price to Soex	0.70
Seeds	0.03
Pay for sector chief	0.03
Transport from farm to collecting station	0.05
Grading, packing	0.14
Cartons	0.14
Depreciation, buildings, taxes and general expenses	0.05
Telephone, telex, water and electricity	0.03
Transport from packing station to airport	0.02
Transit and loading charges	0.05
Total cost	$1.24

Kallu, vegetable producer/wholesaler – Sierra Leone

Kallu inherited his farm from his parents: it consisted of a hectare of swamp land in a small river valley and a rather larger area of hillside on either side. Kallu's parents only cultivated the swamp land, which was regularly flooded in the rainy season and was extremely fertile; they did not use the valley slopes since they were too dry and steep. They had a number of fruit trees and they also grew foreign vegetables such as cauliflower, celery and carrots. The farm is in one of the mountain villages of Nebstown: a favourable location for growing 'European' vegetables, only 10 km from Freetown, the capital of Sierra Leone.

Kallu started on a part-time basis as he also had a good position with the government in the city. He terraced the hillsides above the river valley and nearly quadrupled the total area of the farm. Unlike the local people who depended completely on their holdings for their income, he was able to re-invest the proceeds of the vegetables he sold in improvements of this sort. After two years Kallu realized that what had started more or less as a hobby was growing into a commercial operation. He used to deliver the vegetables to retailers in the town in his own private car, and he found that he could not satisfy their demand. He bought a small hand-pump to irrigate the valley area during the dry season so that he could then grow vegetables through-out the year instead of only in the rainy season like his neighbours. The slopes which he had terraced proved too dry, except during the rainy season, so he invested over $1000 in a pressure pump with a 70 metre head. This allowed him to extend the terraced area further up the hillside. He also invested in shelters to protect his plants from the heavy rains. He built a large storage tank at the top of the hill in order to ensure a regular supply of water.

At this stage Kallu employed two people, but still could not meet all the demands for vegetables. In addition to taking vegetables in the back of his car on his way to work every morning, he made special journeys in the evenings and at weekends which caused heavy wear and tear on his car because of the very bad condition of the road between Nebstown and the city centre. Kallu's retail and wholesale customers in the city continued to press for more vegetables and fruit, and his farm became a regular show place for visitors from at home and abroad. He won a number of prizes at horticultural shows in the city, and individual trade and personal customers started to come to his farm in order to buy on the spot.

Kallu's neighbours were in most cases not so fortunate as he was. Because they had to depend on their farms for their total income, they spent everything they earned and were thus unable to invest in such improvements as terracing or irrigation. Because they were short of money they often had to sell their crops before they were completely ready and since none of them owned a car they had to spend two or three days each week carrying the vegetables to town on foot and selling them to the nearest and most convenient buyer. This not only wasted time which might otherwise have been spent on cultivation, but also meant that they had to accept far lower prices than Kallu for their crops. This was partly the result of their somewhat lower quality: Kallu was able to buy imported seeds in the city, or even on occasion order them himself from abroad, and he also bought fertilizer from the Ministry of Agriculture. Most of the other farmers used local seeds and limited supplies of manure from local poultry farms.

In spite of these difficulties, however, a number of Kallu's neighbours started to copy his methods. A few of them merged their small-holdings, and terraced the sloping land above the river as Kallu had done. They were still unable to improve the marketing of their increased crops, however, and Kallu realized that he had an opportunity both to help his fellow citizens and to increase the scale and the profits of his own enterprise. Most of his customers were retailers: they usually paid Kallu about two-thirds of their final selling price for vegetables. They would for instance pay $1 for a basket of carrots which would be resold for $1.50 or 40 cents for lettuces which would then be sold for 60 cents each. The wholesalers to whom most of the rest of the farmers sold their produce used to work on more or less the same basis. They would pay about 60 cents for carrots which were eventually sold for $1.50, or between 25 and 30 cents for a lettuce for which the final customer paid about 60 cents. Kallu thought that he might be able to pay rather more than the wholesalers did, and, at the same time, make a sufficient margin to cover his own costs. If he could supply more vegetables this would also protect his position since a number of retailers had said that they would like to buy from a single supplier who was able to give them everything they wanted.

Kallu considered the possibility of trying to form a farmer's co-operative so that all the village people would have a say in the management of the marketing of their vegetables, but he decided that since his own operation was far larger than any of the other villagers, and since the record of co-operatives in the country was not good, that he would do better to run it as a private business. The villagers were only too willing to sell their vegetables to him at the prices they would have received from wholesalers in the town, since they were thus spared the necessity of having to spend so much time and effort carrying their vegetables on foot. In a short time Kallu found that he was buying $100 worth or more of vegetables from his neighbours every month and they were more than satisfied with this new outlet for their produce.

Kallu soon found, however, that this had its problems. Some of the retailers noted that the vegetables were not all of the quality they had come to expect. They complained that Kallu's standards were going down and refused to pay the same prices as they had before. Although his neighbours had adopted some of his practices, they could not guarantee the same regular year-round supply as he could. Kallu soon found that it was not enough to ask them to adopt the same methods as he had: he had to spend an increasing amount of time teaching and demonstrating. He bought more seeds than he needed so that he could sell some of them to his suppliers, at no profit to himself.

Reviewing his position over the four years since he had started to keep records of his market gardening activity, he found that his sales had doubled but his profit had been halved: he now had approximately $5000 invested in the enterprise, in addition to the value of the land, and, although he enjoyed his farming, he found that more and more of his time was involved in working with his neighbours and administering the business, which appeared to be neither enjoyable nor profitable. He knew that his neighbours would resist any attempt of his to reduce the prices he paid to them. They would come to regard him as an extortionate middle man, and in any case he had no desire to cause them hardship.

Kallu felt that he was becoming a one-man development agency and he had neither the time, money nor ability to undertake the task properly, but there was no obvious government agency or other way in which both the farmers' and the consumers' needs could be satisfied. Sierra Leone imports several million dollars worth of fresh vegetables every year, including large quantities of those same varieties that Kallu and his neighbours were growing. Most large farmers find it more profitable to produce cash crops for export. In the meantime, large numbers of people are unemployed, marginal land is unused and the drain on scarce foreign exchange continues.

Enebor, rice miller – Nigeria

Enebor had a good job in Benin, the capital of Bendel State in Nigeria, but he hated city life. In 1961 he responded to a government-sponsored 'Farmers' Crusade' and acquired a 2 hectare rice farm in the Illushi Plains on the banks of the River Niger. This is an area of high rice production potential. Illushi is a town with a population of 3000 whose main occupation is farming and fishing. There is a regular weekly market which attracts traders from the north and from nearby states.

For a while he continued to live in Benin and only went to Illushi for brief periods to supervise his farm. As the need for labour increased, the able-bodied members of his family moved to Illushi permanently. Early in the 1970s Enebor finally gave up his city job and moved his whole family to Illushi. He started a small trading enterprise to supplement his income and concentrated on rice growing, processing and marketing. Today, he is one of the most successful rice farmers and processors in the area: his farm has grown from two to 22 hectares and his total enterprise has an annual turnover exceeding $30 000. There is a large and increasing demand for rice in Nigeria. Much of this is presently satisfied by imports, but the unsatisfied demand for home-produced rice provides a good opportunity for people such as Enebor.

Enebor's enterprise includes production, processing and marketing. Table 3.3 summarizes the operation in financial terms: As can be seen from these figures rice production has now become less important. Processing rice, either as a service to others or for sale, is his main operation. Enebor has invested several thousand dollars in equipment, mostly from his own savings and from profits generated by the business itself. He dislikes being in debt and has refused to apply for a loan although he might well be able to obtain one. He has, however, bought some of his equipment on a hire purchase basis, because he is aware of the profitable opportunities open to him. Enebor would like to expand his farm further but has not done so because labour is expensive and it is difficult to obtain seed and fertilizer. He feels that it is less risky to devote his efforts and whatever funds he can mobilize to the expansion of contract milling and sale of rice grown by others.

Enebor clearly has a profitable business which provides employment for several members of his family and a number of other people. He provides an outlet for paddy for a number of farmers and, together with other businessmen, he has played a part in developing the rice industry in Illushi. They now process rice from a number of other districts. 'Illushi' rice is well known as a guarantee of good quality far beyond the immediate neighbourhood.

One of the main reasons for Enebor's success is his own determination and his distaste for city living. He was determined to make a livelihood for himself in Illushi and persisted when a less determined entrepreneur might have switched to some other activity. There is, however, no good road from Illushi. Although the weekly market is well attended no attempt has been made to construct a permanent market. There is no electricity so mills have to be run on petrol or diesel which are expensive. There is no hospital or dispensary and only one primary school. The various rice growers and millers of Illushi would like to organize themselves into a co-operative in order to benefit from

Table 3.3. *Simplified annual operating statement: Enebor, Illushi, Nigeria.*

	$	$	$
Receipts			
Direct sale of rice		13 760	
Parboiling, milling and other services		17 600	
Total receipts			31 360
Expenditure			
Own cultivation costs:			
Tractor hire and farm labour	3 810		
Seeds	100		
Fertilizer	260		
Transport	320		
Land rent	350		
Total cost of own paddy		4 840	
Direct purchase of paddy from other farmers		8 280	
Total cost of paddy		13 120	
Processing costs:			
Manager and skilled operator wages	7 680		
Casual labour wages	2 780		
Rent of premises	480		
Fuel	1 090		
Depreciation	800		
Mechanical repairs	450		
Total processing costs		13 280	
Total expenditure			26 400
Net surplus			$4 960

government credit schemes, small-scale industry training and other programmes, but there has been little progress in turning their ideas into reality. In general, Illushi has been bypassed by government programmes. There has been little investment in infrastructure and various schemes to support small industry are concentrated in more accessible places.

Al Haji, marketing board buying agent – Cameroun

Until 1976 Al Haji was a traditional farmer in the neighbourhood of Maroua. Each year he harvested between 400 and 500 sacks of rainy season sorghum. He always had difficulty in selling the surplus before it started to go mouldy and when there was a good harvest he was unable to sell all his crop in the market in Dargala. He could take it to Maroua about 25 km away but sometimes it was difficult to sell it all there. To avoid total loss he had to lend his grain to other producers in return for similar amounts of their own crops which they promised to give him when they next harvested. This was not a satisfactory arrangement, but there was no alternative at the time.

Wide variations in rainfall were responsible for periodic surplus and famine in Cameroun. After the major drought of 1970 the government set up a Cereal Marketing Board to cover the Northern Province. Its principal task was to stabilize the price of millet and sorghum which, together with maize, form the staple diet of the people. It purchased after the harvest when prices were low, held stocks for several months and then sold during the 'hungry period', between the months of July and October. The Cereal Board had no monopoly and worked in a free market.

When the Cereal Board started to operate in Diemare District, in 1976, Al Haji, like the other farmers, was immediately able to sell all his surplus sorghum. He was delighted when the Controller of the Board asked him to be one of the Board's buying agents. In 1978 he bought 3386 sacks and earned $1608 commission; in 1979 6600 sacks, earning him $3927. The commission per sack was increased from $47\frac{1}{2}$ cents to $59\frac{1}{2}$ cents a bag in 1979 to allow for higher costs. In 1980 Al Haji's purchases fell to 2866 sacks because the drought that year reduced farmers' marketable surplus.

Al Haji buys from the six public weekly markets in his area and also direct from farmers in villages up to 15 km away. The Cereal Board advances finance and provides empty sacks. Together with 15 workers

Al Haji issues these sacks to the farmers, collects them when they are filled and submits them to the market inspectors employed by the Cereal Board. They verify that the sacks contain the equivalent of 100 cups of a size specified by the Board, and check the quality. The inspector then informs the buying office of the Board of the pick-up point. From the moment the sacks of sorghum are collected by the Board's vehicle they become its property; up to that point Al Haji is responsible. The grain that Al Haji buys direct from farmers is priced at the rate current on the day it is checked by the market inspector; that which is bought in the village markets is priced at the rate current on the day of the market. The total value is debited to Al Haji's account with the Cereal Board.

Al Haji has had to make certain investments for his buying operation. He has five motorcycles, one for himself and four for his workers, five bicycles for his other workers and six donkeys to carry the grain. He is also training two oxen to pull a cart which he hopes to buy shortly. The total value of all this property, not including the oxen, is approximately $2000. This is a substantial investment for a farmer such as Al Haji but he considers it worthwhile. His estimated receipts and expenditure which are shown in Table 3.4 relate to 1980 which was a very bad year

Table 3.4. *Profit and loss account, Al Haji, 1980.*

	$	$
Receipts		
Commission from Cereal Board		
2 886 sacks at 59½ cents		1 710
Expenses		
Commission paid to workers		
2 866 sacks at 24 cents	690	
Storage expenses		
1 433 sacks at 18 cents	260	
Fee to donkey driver		
1 433 sacks at 9½ cents	140	
Binders for tying sacks	70	
Needles, measuring cups, books for accounts	20	
Motorcycle operating costs	190	
Total		1 370
Net profit		$340

with little surplus grain on offer to the Board. The Board agreed to an increased commission of 63 cents per sack for 1981.

Al Haji pays his workers a commission of 24 cents, which they divide among themselves, for each sack they collect. The donkey driver receives 9½ cents for each sack he carries. The donkeys also have to be fed: each animal eats about one bucket of millet every three days, but this is not included as an expense because there is always some damaged millet available.

The two major problems faced by Al Haji are delays in receipt of funds to purchase grain and delays in provision of transport to take the grain to the Board's warehouses. When Al Haji has no funds the farmers are unwilling to sell to him. If the Board does not send transport when the inspectors order it the grain may be stolen or damaged and Al Haji will be held responsible. These two problems are related because the Board insists on issuing funds in regular small amounts and will not make a further advance until the previous one has been accounted for; the delays in transport therefore delay the issuing of further advances.

Al Haji and four other buying agents play a vital role in the supply of cereals to the Cereal Board in Diemare District. In 1979, Al Haji provided about one-third of the sorghum the Board received. He has good contacts; his house provides free storage; he is only paid for what he does. Farmers are pleased with the service they receive because his measuring cups are accurate, he works fast without making them wait and he is polite. Al Haji is a well-known personality in his own village and has good long-standing relationships with a number of farmers in the neighbouring villages. He is illiterate in French, but he can write in Arabic. He is energetic, hard-working, honest and trustworthy.

The farmers who sell him their millet know that he will pay them the correct market price, saving them a long journey to sell it elsewhere; they know that he will always collect filled sacks as quickly as he can, and that he will pay them promptly. As a result, he has a large number of permanent suppliers on whom he can rely every year if they have a good harvest.

Wimalajeewa, market commission agent – Colombo

Wimalajeewa is one of 670 commission agents operating in the Colombo wholesale market. He learned the trade working with his uncle from when he was 18 until he was 21 in 1942. He then started business on his own with a total capital of $1. He rented a stall for 30 cents a month from the Municipal Council and started up as a commission agent for golden melons. Later he diversified into other vegetables. It is now very difficult to obtain such a stall. A new entrant has to pay up to $10 000 as key money. Wimalajeewa, however, still only pays $5 a month because he has been there so long.

Wimalajeewa takes no legal title to the vegetables which he sells. He receives them from growers and then sells them to wholesalers for the highest price he can. He retains 10 per cent of the price he gets and remits the balance to the grower, less transport charges from the producing area which he has paid on receipt of the produce.

Wimalajeewa has a very open market position. He cannot make contracts with wholesalers since he does not know what produce he is going to receive. Nor are his customers bound to purchase from him regularly. He employs labour to unload and store his vegetables. The wholesalers who buy through him pay for transport costs from his premises to their own.

He deals in a range of vegetables produced on farms of up to one hectare with most of the work done by family labour. Some of these growers sell their produce at local markets, while others work through agents in Colombo. About 200 farmers regularly send their produce to Wimalajeewa. They harvest it in the evening, pack it without much cleaning in gunny bags or wooden crates and take it to the main road to be picked up by any lorry driver who is transporting vegetables to Colombo. They obtain a receipt from the driver and send a delivery note stating the quantity that has been loaded. Wimalajeewa receives the packages at the market early in the morning; he weighs them, enters the quantity in his register and then pays the lorry hire which is deducted from the amount due to the supplier. Transport costs vary between 8 cents and 12 cents per case per hundred km. Wimalajeewa sells the produce the same day and sends the grower a note by mail stating the quantity received, the selling price, his commission, the transport charge and the balance payable. This amount is then sent to the grower by cheque the same or the following day. If the goods are not sold on the day they are received, Wimalajeewa sends a deposit to the producer and then clears the account once the goods have been sold.

Wimalajeewa hopes that one of his two sons will carry on the business when he retires although he realizes that there are many problems. During the last two years he has had to write off about $1500 in advances made to farmers which were never cleared by deliveries of vegetables. It is traditional in this trade to make advances for weddings, funerals, family emergencies, or to cover the advance costs of cultivation. On occasion, growers fail to redeem these advances, particularly when they find that an alternative commission agent can obtain a higher selling price. Growers tend to justify this by saying that commission agents do not actually declare the real price which they obtain for their vegetables, so they actually make a higher profit than the stated 10 per cent commission. It is difficult to substantiate this because transactions at the wholesale market are usually by word of mouth. The majority of growers nevertheless prefer to sell their produce through commission agents such as Wimalajeewa, rather than through co-operative purchasing agents or through the Government Marketing Department.

In a typical transaction, Wimalajeewa sold a consignment of cabbage for $3 and deducted 30 cents for his own commission and a further 70 cents for the cost of transport from the farmer to Colombo. The farmer thus received $2 for the consignment. The same cabbages, when cleaned, were sold retail for $12. The wholesalers explain differences of this magnitude by pointing to the high cost of transport from the Colombo market to the wholesaler's premises and thence to the retail shop. There is also a cleaning loss of between 5 per cent and 25 per cent depending on the vegetable; it is highest in the case of cabbages. Retailers, and wholesalers to a lesser extent, also suffer substantial loss through deterioration in vegetables which are held for more than a day.

In spite of their complaints, farmers continue to sell through the traditional commission agents, wholesalers and retailers. Though the proportion of the final selling price which they receive may be small, wholesale market prices appear to have kept up with inflation. As consumer demand increases and farmers benefit from improved varieties and better cultivation methods, their incomes have also substantially improved.

Mafandala, produce buyer and retailer – Zaire

Mafandala started his trading business by opening a shop in Lusanga in 1966. At that time he only had one Chevrolet pick-up. He bought a lorry with the profits he made from trading in beer. With this he was able to start a two-way trade with local farmers. He would buy crops such as maize, cassava, groundnuts and gourds from them and transport them to the town, at the same time taking other foods and farm inputs from his shop out to the farmers. As his business developed, Mafandala opened shops in Kikanzi, Bisiri and Bulungu. Bulungu is an important commercial centre because of its location on the River Kwilu which provides access to a large area of good agricultural land.

Between 1972 and 1976 Mafandala's main business was trading in coffee, which was at that time very profitable. His business grew rapidly, because he was able to make annual sales contracts with large commercial enterprises and obtain loans on their security. The business also benefited very much from his success in obtaining regular quotas of beer from a large brewery in Kinshasa. He sold between 300 and 400 cases a week, making a very good profit on each case.

Currency reform

In December 1979, to improve the country's fiscal situation, the Government devalued all the currency which was in people's hands rather than in the banking system. Many up-country traders lost their capital through this move, since they had it in ready cash to be able to pay farmers during the harvest. The effect of this Government edict was more serious in rural areas than in the cities because there was no effective banking system in the countryside.

To Mafandala this was a severe blow, but not fatal. He had his stocks of foodstuffs. However, the regional authorities compelled all traders such as Mafandala to keep their shops open and to allow the local population to buy with old notes which could be exchanged on the black market for 40 per cent of their original value. This was a further blow, but the only businesses which survived were those like Mafandala's who had large stocks of agricultural produce at the time of devaluation. Slowly they were able to recover their capital.

Difficult operating conditions

Mafandala now buys produce from an area of about 100 km around Bulungu. He trades in maize, cassava, groundnuts and palmeto, selling them in the market in Kinshasa where the highest prices can be obtained. Mafandala transports his produce to Kinshasa in four of his own lorries, as well as hiring transport locally. In his buying areas he supplies farmers with soap, tinned sardines, salt, petrol, textiles and other goods. Often where local roads are impassable because the Roads Department fails to maintain them, Mafandala has had to take over responsibility for their up-keep in order to be able to collect crops from his farmers. His business is also suffering from increased transport costs. An 8-ton Fiat lorry which could have been bought in 1978 for $1800, now costs $90 000. Sometimes he has to pay $700 for 200 litres of diesel on the black market when the official price is $350.

Devaluation of the currency caused a general decline in economic activity, and Mafandala's business is lucky to be operating at half its previous scale. Overheads now have more impact on the business since the warehouse and the lorries cannot be fully utilized.

When the business was operating at its peak, and coffee business was very profitable, Mafandala could obtain credit on the basis of contracts with large customers. This is no longer possible and he is applying for a loan from a bank to provide the necessary working capital. Financial results appear to justify such a step. His profits as a percentage of turnover rose by 23 per cent in 1978, and in 1979 by 46 per cent. In 1980, however, they fell by 41 per cent because of the effects of the devaluation. It is important, when evaluating these figures, to remember that the prices of agricultural products rose substantially during the 1980 season.

Total sales increased by about 80 per cent over the three-year period with produce amounting to 74 per cent of the total. Some of this was the result of inflation, but Mafandala succeeded in keeping management costs at about the same figure. Stock levels were kept at a minimum, and the financial affairs of the business were generally well conducted.

The results, however, were achieved without any loan finance and do not therefore allow for the very high interest rates which would have had to be paid if a loan had been obtained. Until now Mafandala's business has been operated largely on its own funds and he has not even taken account of his own salary when calculating his profits; Some allowance for this will have to be deducted from the net profits in order to give an accurate picture to the bank.

Matlhaku, butcher – Botswana

Ismail Matlhaku was a labourer in Orlando West, one of the urban townships outside Johannesburg in South Africa. To supplement his income he sold soft goods such as stockings, wash-cloths and handkerchiefs to the people in his neighbourhood. After five years he had accumulated sufficient savings to open a small vegetable shop in the town and buy an old truck for selling vegetables from door to door.

In 1963 he emigrated to Botswana for political reasons. He took his

old truck and a ten-seater minibus which he had bought in Johannesburg but he left his family behind in South Africa and looked around
for a source of income and a new home in Botswana. He saw that the
local transport facilities were very inadequate, so he obtained a licence
to run a taxi service with his minibus between Pilane railway station
and what was then the village of Gaborone. The hours were very
difficult; he was working until 3.00 a.m. every morning so he decided
to look for some other opportunity which would allow more reasonable
working hours and decided to go into the butchery business. He chose
butchery partly because cattle were readily available in Botswana
whereas nearly all the groceries sold in Botswana came from South
Africa. In the butchery business he would not need to do any business
with South Africa.

Matlhaku started by buying cattle, sheep and goats and re-selling
them to the Botswana Meat Commission. He bought the animals fairly
cheaply, wherever he could, and fattened them up on rented grazing
land, with some supplementary feeding. After a few months Matlhaku
felt ready to go into the butchery business proper. He obtained a
licence to open a butchery in Mochudi, the town where he lived, some
43 km away from Gaborone. He put up a simple building near his
house using bricks which he had obtained free from the brickworks in
return for transporting ash with his truck. By this time his wife had
joined him from South Africa and they ran the business between them.
Initially Matlhaku employed somebody else to do the slaughtering, but
he soon learnt how to do it himself and was able to operate on his own.

That same year Gaborone was selected as the capital of independent
Botswana which led to an enormous influx of civil servants and
construction workers and to a great increase in demand for meat.
Matlhaku bought a roof-carrier for his minibus so that he could
transport meat from Mochudi to Gaborone on the top, and could still
use the inside of the vehicle for his taxi business. He was now
employing a driver for the minibus and two boys to deliver meat from
house to house, and the business expanded rapidly.

Matlhaku soon found it necessary to acquire a small depot in
Gaborone to hold the meat ready for delivery to individual households.
The butchers who were already operating in Gaborone, most of whom
were whites, resented an outsider from Mochudi interfering in their
territory, they suggested that it was unhygienic to transport meat from
Mochudi to Gaborone before selling it. Matlhaku resisted their pro-
tests, and carried on his business.

In 1969 a good friend of Matlhaku offered him the rental of a large shop in Gaborone to convert into a butchery. Through an acquaintance on the town council he managed to continue operating on a provisional basis while he was obtaining the necessary equipment.

Matlhaku continued to slaughter animals himself for the butchery in Mochudi, but arranged for those which were to be sold in Gaborone to be slaughtered in the official abattoir in the town. By 1976 Matlhaku was employing two people in Mochudi and five in Gaborone. He had now saved enough money to open a further butchery in Gaborone and employed three more people to operate it. However, as he was now in his late 50s, he found it exhausting to manage all three butcheries. In 1980 he sold the new butchery in Gaborone and delegated the management of the other to his wife; he continued to run the slaughtering and butchery business in Mochudi.

Matlhaku realized a substantial amount of money from the third butchery and invested this in a bar and bottle store in Artesia, another town 44 km from Mochudi.

Mochudi remains the main base for Matlhaku's butchery operations because that is where he keeps his cattle for slaughter or indeed for sale if he needs ready cash. He buys live animals direct from farmers, and butchered meat from the Botswana Meat Commission. When he buys from the Commission he has to pay rather higher prices; when he buys privately he can sometimes get very good bargains. He normally strikes an average when fixing his selling prices and the resulting competitive prices have attracted a very loyal group of customers. He normally sells the meat from about eight cattle per week; during Easter and Christmas, and at the end of the month, the number goes up to twelve or more.

Matlhaku is generally satisfied with his business. He has no ambition to expand because he wishes to retain control of the total operation and does not wish to work too hard now that he is becoming old. He has turned some of his attention to farming and has planted maize on land allocated to him by the local Chief in Mochudi when he arrived in Botswana in 1963. He sold about $2500 worth of maize in 1981 and wants to increase the area planted.

In order to finance this increased agricultural activity Matlhaku tried to borrow $20 000 from the bank with whom he has been doing business for the last 17 years. He furnished the following profit and loss account (Table 3.5) and balance sheet (Table 3.6). Matlhaku offered his butchery buildings and his land as collateral, to a total value of around

Table 3.5. *Profit and loss account for year ending 30th June, 1980:*
Matlhaku's butcheries.

	$
Gross income	
Sales proceeds	115 710
Cost of sales	72 670
Gross margin	43 040
Rent received from buildings	2 020
Total	45 060
Expenses	
Wages	6 660
Motor vehicle expenses	7 320
Electricity, gas and water	2 770
Repairs and maintenance	2 990
Hire purchase interest	1 340
Licences and insurance	860
Depreciation	4 850
Bank charges and interest	1 490
Accounting fees	600
Telephone and postage	400
Stationery	10
Miscellaneous	2 490
Rent and rates	1 710
Total	33 490
Net income	11 570

Table 3.6. *Balance sheet as at 30th June, 1980: Matlhaku's butcheries.*

	$		$	$
Assets		*Liabilities*		
Cash	nil	Bank overdraft		130
Accounts receivable	1 110	Accounts payable		1 560
Stocks	100	Hire purchase owed		3 380
Furniture and		Owner's investment	40 870	
equipment	10 390	Profit for year	11 560	
Vehicles	14 790		52 430	
Buildings	16 580	Less owner's		
		withdrawals	14 530	
		Net owner's equity		37 900
Total assets	$42 970	Total liabilities		$42 970

$75 000, but the bank would still only lend him $10 000. However, a competing commercial bank has offered him the money he needs. Matlhaku feels that he will be able to obtain the funds either from this bank or from his own bank when it becomes aware of what the competition has offered.

The 'market kings' of Sudan

Apart from areas of irrigated farming along the Nile agriculture in Sudan is very extensive. Transport has been a major constraint on development. Construction of a rail link to Port Sudan provided an initial impetus, but its capacity is limited. Earth feeder roads can be impassable for two to three months a year and tarmac surfaces are extending only slowly.

Khartoum, Khartoum North and Omdurman, known as the Three Cities, constitute the main urban agglomeration with a total population of over one million. They have become increasingly dependent on food supplies brought from a distance.

The major development in Sudan has been the Gezira Irrigation Scheme for the production of long staple cotton, on which the country's economy still depends. Little publicized has been the simultaneous development of the commerce of the country by the members of certain families – the 'market kings' of Sudan. It was they who were the early entrepreneurs. They introduced a money economy linking subsistence farming with the outside world and out of their profits they built factories and established transport companies and other economic activities.

The commercial structure they created hinged on caste and kinship. In one family, where the original entrepreneur established his business in the late 19th century, there are over 150 living descendants. None are less than second or third cousins and many are related more closely because of the Northern Sudanese custom of cousin marriages. The 'market kings' are not so much individual entrepreneurs as commercial houses owned and usually managed at major and even minor levels by people related to one another. They are usually descended from one common ancestor – the original founder of the business. These organizations are very much a closed shop. They constitute a group of relatives with a strong sense of loyalty and family discipline who are not going to be tempted away to work for other employers. Nor is there much leakage of vital trading information. The kinship network helps to ensure loyalty and monitor deviations from expected norms of behaviour.

Although these family businesses are based in Khartoum and Port Sudan, their activities extend over the whole country. Several, indeed, started in the rural areas. They distributed imported goods, such as tea, sugar, china and gaily coloured women's clothes. It was the offering of these 'goodies' that stimulated the counter-offering of agricultural produce; a cause and effect relationship.

The founder of two of the major Sudanese trading houses – the Aboulela group of companies and the Abel Moneim Mahomed Co. – came from Upper Egypt in the 1870s. He began as a pedlar in the Gezira province riding on a donkey from village to village selling china cups. His personal security was not assured and he had several narrow escapes from death at the hands of robbers. On one occasion he is said to have thrust some coins down a donkey's throat at the sign of armed robbers and then killed the unfortunate donkey afterwards to retrieve the money! The main product then that justified transport over long distances was gum arabic – the crystallized sap of two species of acacia which grow wild in the semi-arid savannah. Primary assembly of gum is still by camel transport to some extent.

Gedaref was the point of origin for another set of traders – Copts from Egypt. It was linked to both Khartoum and Port Sudan by rail (see Fig. 3.1) and became the administrative headquarters for the Butana area. These immigrants came originally to serve the expatriate British, but found customers also among the local people. To pay for consumer goods on offer at their shops, subsistence farmers were encouraged to sell gum arabic, sorghum and sesame. Grain and oilseed production of the Butana date from these beginnings with the great grain silo at Gedaref which dominates the skyline.

In 1966 five major merchants marketed sorghum at Gedaref. A more detailed survey in 1975 showed these families still to be paramount. Several by then had moved into mechanized farming, with up to 12 000 hectares each. Their success contrasts with the fiasco of the large mechanized state farms for sorghum and sesame production. The Gedaref farmers were able to keep their farms going when low yields in dry years made them financially unviable.

These private sector concerns also made up a close-knit society with inter-marriage between leading families of the same religion. That they were Christian did not isolate them from the local community or from the mercantile classes elsewhere in Sudan. The Moslem Egyptians in

Sudan tended to work with Copts from Egypt because they felt a cultural 'kinship'. This common Egyptian background also helped in the export of produce to Egypt and the Arabian peninsula.

The Port Sudan firm of Mohamed Sayed El Barbary is another example of a kinship network in foreign trading. The founder, who only died a few years ago, came to Port Sudan from Aden as a young

Fig. 3.1. Sudan: railways and surfaced roads.

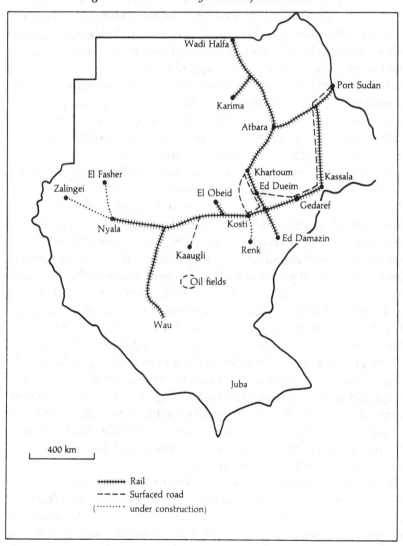

man to work in the docks. He soon started trading on his own and, by the 1920s, had built up a sufficient private fortune to buy the Sudan Plantation Syndicate's Zeidab Scheme for irrigated cotton production in the Northern Province. He is said to have been the only Sudanese (apart from religious leaders) who has been able to mobilize sufficient capital to do so. He had quickly made use of the opportunities offered by the pacification of the country and the construction of a rail network to a deep water port, originally exporting ivory and gum arabic.

By the 1970s, El Barbary had an 80 to 90 per cent share of sorghum exports. The firm worked closely with the large merchant/farmers of Gedaref. When faced by shipping delays at Jeddah due to congestion in the harbour, the firm chartered small dhows to carry sorghum across the Red Sea, a day's journey with a good wind, and unload it up small creeks along the Saudi coast. Meanwhile, rival shipments from the United States were held up for weeks in large freighters queuing to unload at Jeddah. These shippers eventually used helicopters to unload but their bags split and a lot of the grain was lost in the sea. The traditional technology of the dhow was more successful.

The livestock and meat trade provides another example of the willingness of entrepreneurs to venture into the wilds. Through their 'Egyptian connection' they built up a major export of cattle and sheep from nomadic grazing. Marketing began with itinerant pedlars waiting for the nomads at wells on their regular migratory routes where mature animals were exchanged for tea, coffee, sugar, salt, spices and textiles. They also pledged young stock in advance of delivery: they had to be fully grown to withstand the trek to Omdurman which could take up to three months. Some of the animals would be sold for slaughter in the Three Towns, those for export quarantined and then sent by train to Port Sudan. Cattle went mainly to Egypt, sheep to Saudi Arabia.

The predominance of five family firms in this trade goes back to the sharp increase in demand for meat during the Second World War. Large allied armies were stationed in Egypt. These firms obtained contracts with the British authorities for the supply of cattle to the armed forces. They could then obtain commercial bank credit to finance their purchase. They were shrewd and had the business acumen to exploit these opportunities.

Other 'market kings' include the almost legendary Sayed Haggar who was a merchant adventurer in the upper reaches of the White Nile and along the Congo River. He began by buying crocodile skins, then moved into dried and pickled wet fish for export, largely to Egypt.

Osman Saleh and Sons pioneered the oil seeds trade and the oil-crushing industry. At one point the company was nationalized and incorporated into a state Oil Seeds Marketing Company but it incurred heavy losses through inefficient management and the business was returned to its original owners.

Various Greek families also built up trading networks in the Sudan. However, independence brought a surge of feeling for 'national capitalists'. The Nimery Government proclaimed that it was an alliance of workers, soldiers and national capitalists. The Sudan market kings were in a good position to take over import and export agencies from former British companies.

The first stage in the climb to fortune of the market kings was to accumulate the initial capital necessary to become creditworthy in the eyes of a bank. This they accomplished through hard work, parsimonious habits and the exploitation of every opportunity that came their way. One favourable factor was the great expansion of the Khartoum market in the Second World War. The foundation of several Khartoum fortunes were laid in contracting to the British army. There were also opportunities for lucrative smuggling.

Economies of scale

Income and expenditure accounts for these firms are unfortunately not available so it is not known what rate of return was obtained on their capital. That they were relatively efficient is evident from the experience with nationalization of the oil seeds business. The parastatal company incurred much greater costs through over-manning and failure to meet delivery dates.

A basic issue for development is the tendency, as in Sudan, for the wholesaling of agricultural produce to become concentrated in the hands of relatively few firms because of economies of scale. Large market shares give corresponding bargaining strength to the firms concerned who are then in a position to adjust prices either to earn high profits or to pass on high costs, or both. Lack of competition may enable them to underpay producers for their output or/and overcharge them for supplies. If carried too far, such action can hold back development because farmers are denied the incentives and resources to increase their productivity.

In the Sudan, capital accumulated by these dominant family firms seems to have been ploughed back into the family enterprise. They then diversified into agricultural and industrial production which added to the wealth of the country and created additional employment. Because these production activities were integrated with their suppliers and their markets through an associated marketing enterprise they constituted an effective economic development.

Attributes, advantages and support needs

The preceding cases provide a body of experience from which we can deduce characteristics of indigenous entrepreneurs who have contributed significantly to their success. There are features of these operations that make them advantageous to an economy and areas in which government support is important.

Attributes

Personal initiative is characteristic of an individual who sets up and runs a marketing enterprise. He must be ready to accept risks and able to look ahead and see what margin will remain after buying at one price, selling at another and covering the costs involved. Mary Jane of Dominica took the risk of carrying produce by air to an island using a language other than her own.

Ability to take quick decisions is a key attribute of the private enterprise. The responsible operator frequently has to assess the quality of produce on offer by eye, recognize where he can sell it again, and for how much, and pay cash to clinch a transaction. When access to the harbour at Jeddah was blocked, the Sudan sorghum exporter promptly chartered local shipping able to use other ports.

Independence of spirit and persistence are also essential qualities of the individual marketing entrepreneur. This is demonstrated by Matlhaku's determination to find a new niche for himself by moving on from one activity to another, and by Enebor's determination to succeed in Illushi.

Willingness to work hard, for long or irregular hours is implicit in the private marketing enterprise. It can rarely count on protection and an easy life. Even when the operator is in a position to employ staff to undertake the more routine tasks, he must be ready to fill gaps himself, to deal with problems whenever they occur, putting this obligation before family or leisure priorities. During the vegetable export season, M. Diallo of Soex worked 16 hours per day including Saturdays and Sundays. Like many migrants, Matlhaku worked extraordinarily long hours in his early years. Kallu in Sierra Leone was able to build up a vegetable marketing business in addition to holding a government position only because he was prepared to work twice the hours of his civil service colleagues.

Relevant experience Petty retail trading has often been the recourse of people who could find no other employment. It requires only a small initial capital or credit, plus a willingness to stand in the street and sell. Most 'successful' private marketing operators have, however, some experience that enables them to judge what to expect from a particular course of action and gives them the confidence to go ahead.

An agricultural background is common in marketing entrepreneurs who deal with farmers and their products. Enebor, the rice miller in Nigeria, started with farming and then moved into processing and marketing the crops of other farmers. Knowledge of how crops are grown and mature is a valuable precursor to buying them from other people. There can be internal economies in combining production and marketing. Assurance of some of one's own supplies helps in operating processing equipment to the best advantage and in maintaining a regular buyer clientele. Fuller use may also be made of available family labour.

As a farmer himself, Al Haji, the marketing board buying agent, knew how fellow farmers thought and behaved and was aware of their problems. This helped him orient his service to their needs. He also started with the advantage of being trusted because he came from the same background and spoke the same language.

At the same time *experience outside the immediate environment* is one of the characteristics of the successful trader. A key factor is being able to see beyond the horizons of the family or the village and open up channels to a distant market, or introduce technology and procedures from somewhere else into his home area. Enebor the rice miller had worked in Benin. Matlhaku, who set up the retail butcheries in Botswana, had migrated there from Johannesburg and could dispassionately examine the business opportunities available. He had also previously been in various other lines of business. Each bore some relation to the one that had gone before: the experience and capital that he had accumulated could be applied to his latest venture.

A post in government can be a very useful base for innovation in marketing and related activities. Because of their access to information and knowledge of licencing, bank lending and other strategic procedures, government officials in developing countries have often been a main source of indigenous business initiatives. Kallu's venture shows how somebody with a regular income from elsewhere can innovate. He was able to obtain improved seeds and fertilizer from the Ministry of Agriculture whilst other farmers used less satisfactory materials. He could plough the extra income he made back into the business.

In the establishment and growth of a marketing enterprise, operating over a distance and in several places, *kinship ties* can be a major factor. Where communications are difficult and the loyalties of paid staff divided, a ceiling on the distance over which an enterprise can operate is often set by the physical capacity of the operator. Matlhaku of Botswana was able to manage more than one location by working with his wife.

It is a characteristic of the market women who organize the bulk of the food supplies of West African cities that they have a link with a relative in the producing area where supplies are purchased, and with another in the urban centre where they are sold. The loyalty inspired by family ties extending to cousins was one of the great strengths of the 'market kings' of Sudan. Family members could be counted upon to reinforce each other in the face of competition and to conform strictly to the responsibilities delegated to them. This permitted extension of the field of operation far beyond the limits of personal supervision possible by the head of the business.

Advantages of the private enterprise for an economy

The intrinsic advantage of allowing scope for private marketing enterprise is that it is set up and carries on without any burden on the government. It adapts to changing circumstances and opportunities and generally, provides the services for which it is suited at lower cost than available alternatives.

Low operating costs Knowing that every penny saved on unnecessary expense is additional income to the operator has a crucial impact on the attitude to costs of the individual enterprise. It is unlikely that a public or co-operatively-owned business of the scale of Matlhaku's would be able to keep its expenditure on stationery down to $10 a year. While a minor item, it demonstrates the advantages of combining ownership, decision-making and actual operation. This minimizes the need for records, reports and formal communications.

Al Haji was employed as buying agent by the marketing board in Cameroun because he could provide this service cheaply. He was already based in the area and this new responsibility was a supplement to an existing business. In Kenya the cost of a direct marketing board buying agency for maize was estimated at 46 cents per sack in 1980–83 and of one operated by a co-operative 50 cents; 32 cents was the allowance to private buying agents.

Economy in use of equipment and management In many rural areas, private entrepreneurs provide two-way services to the farming community. Mafandala, in Zaire, minimizes both transport and administration. He buys consumer goods in Kinshasa which the rural people need. With his own vehicle he makes them available to farmers where they live. He purchases farmers' crops setting the cost of the supplies provided against payment for the crops and transports the farm produce back to his own town and on to Kinshasa. Such a trade maximizes returns to the transport vehicle and the capital and management involved. Each of the various services provided is maintained at much lower cost than if they were undertaken independently.

In the absence of competition, combined operations of this sort can open the way to exploitation. Because this *may* occur does not mean that it is the general rule. There was no evidence that Mafandala was acting in this way. His business provided a valuable service in a situation where much of the commercial infrastructure, including a reliable currency, transport facilities and fuel, was lacking or ineffective.

Adaptability An intrinsic advantage of the individually operated business is simplicity of decision-making. The man who runs the business decides on the spot how best to deal with a particular situation. This is especially important where operating conditions are uncertain. Mafandala used his personal contacts to obtain quotas of beer which was in short supply. He put the profits from beer into his agricultural service and marketing activities. He took on road maintenance, because he saw it was necessary if his business was to proceed. He obtained fuel through the black market. Adherence to the law could have meant that farmers had neither a market for their crops nor access to the staples and supplies they needed. By trading, possibly illegally, in currency notes devalued abruptly, Mafandala was able to survive and to continue to serve his customers. Thus he both filled the gaps left by the public services and reduced the impact on the rural community of a damaging government edict. One of the greatest attributes of the individual marketing entrepreneur is ability to improvise and find his own practical solutions to problems.

Support requirements

It is sometimes said that the less interference in private marketing operations there is from government the better; and that from the trader's point of view 'no government is good government'. This is a half-truth. Lack of basic infrastructure and support facilities handicaps the private enterprise as much as any other, making its operations more time-consuming, expensive and risky. Transport costs double and triple when truck movements are over unpredictable earth roads as against reliable tarmac. When means of information and communication are lacking, the prices offered for produce will be lower to allow for uncertain conditions of resale.

There are many areas where governments can support indigenous private enterprise and enable it to contribute more effectively to the growth of an economy.

Provision of the basic infrastructure is the first requirement of government together with stable conditions of sale. With these the marketing enterprise is continually handicapped in moving goods from where they are produced to where they are most wanted for consumption. This begins with the construction of roads, railways, ports for general use; and maintaining them in good working order. A legal framework of protection to buyers and sellers against fraud and for enforcement of contracts is a general requirement of ongoing marketing operations together with effective protection against armed robbery and extortionate local demands as a condition of movement.

Access to motor transport is crucial for the agricultural marketing enterprise as is priority in the allocation of foreign exchange to imports of commercial transport vehicles, and their spare parts. The difficulties faced by Mafandala and his fellow traders in Zaire gave rise to an F.A.O. marketing project. This focussed on helping a government agency ensure that traders' vehicles setting out to rural areas to bring back produce could obtain the repairs, spare parts, fuel, etc. they needed. Central and local government can help by organizing special services for marketing that may not be feasible for an individual enterprise – public carriers, for example, following regular routes through rural areas.

Provision of public market facilities is another responsibility best handled by central or local government. Enebor in Nigeria and Al Haji, the marketing board buying agent in Cameroun, operated daily in public markets lacking paving and roofing against mud and rain. The wholesale produce market in Colombo where Wimalajeewa has his commission agency has been plagued for years by difficulties in transport access and congestion, pending a government decision to transfer it to a more convenient site.

Provision of improved market facilities typically includes paved sites and access roads, covered stalls, overnight storage, sanitary, washing, and childcare facilities where there are many women traders, against payment of a fee.

Access to finance Without some working capital or credit, marketing will be restricted to farmers selling their own production to consumers. Access to capital has been so strategic in West Africa, for example, that many farmers have used the proceeds of cocoa sales to buy some other seasonal produce and benefit from the returns from holding it to sell again.

The critical step in the growth of the family marketing enterprises in Sudan was accumulation of enough capital to become creditworthy with a commercial bank. This was done by hard work, parsimonious living and profiting from opportunities as they appeared. $125 borrowed under a youth development programme was the starting point for Mary Jane, the young export trader of Dominica. Assuring access to credit is one of the ways in which governments can support the growth of private marketing enterprises and competition between them.

A favourable and consistent policy position of government is essential. In some socialist economies private wholesaling is officially illegal. If it happens at all it is because local authorities close their eyes where it provides useful service not available through the official system. In various mixed economies private traders incur unnecessary costs to overcome harassment by local officials; they face the risks of arbitrary seizures of stocks and unrealistic price fixing.

Clear government recognition is needed that private enterprises have a major role in an overall marketing system even if other kinds of enterprise prevail in some parts of it. The next step is to establish a support point in an appropriate government department. It should be a communications link with the traders for the discussion of policies concerning them and for organizing specific programmes of assistance.

Training Most of the entrepreneurs featured in this chapter operated on the basis of experience and natural ability. Wimalajeewa was apprenticed with an uncle for three years before starting up on his own. Undoubtedly this was the most relevant training available to him. This does not mean, however, as many sons of farmers and of entrepreneurs in other fields have appreciated, that access to more systematic training would not be beneficial. In marketing, as in other fields of economic activity, training should be provided as a public responsibility for operators and staff of private and other enterprises. In Brazil this is provided by COBAL, a government company responsible for wholesale markets. It should include:

– training in simple marketing management, accounting, sales methods, etc;
– demonstration of improved product handling and packing methods, equipment and materials;
– training in transport, storage, refrigeration, quality control and other relevant management and technology.

Issues for discussion

1. What is the status of private indigenous enterprise in your country? In which lines of commodity marketing is it active?
2. What are the popular attitudes governing entry into private trading? Are there some groups of people in your country who favour this and others who are negative towards it? What are the bases of these attitudes?
3. Prepare marketing profiles of two indigenous enterprises trading in food and agricultural products. How did these businesses originate? What are their main lines of activity? Assess their operating margins on the basis of prevailing prices at producer, wholesale and retail levels; then check these against any other information you can find including data from the enterprise itself.
4. In the marketing channels where indigenous enterprise participates, what would you say was the degree of competition? On what factors do you base your assessment?
5. In your country, are indigenous enterprises like that of the hawker in Dominica and Soex in Senegal engaged in export marketing? What are the constraints on such operations? Can you identify opportunities for them if these constraints were removed?
6. Is two-way rural trading, i.e. sale of farm inputs and consumer goods and purchase of farm output, by indigenous enterprise, common in your country? What factors govern its existence and possible expansion?
7. Are there commission agents like Wimalajeewa on wholesale markets in your country? What are the main factors governing their operations and their number? If such enterprises do not exist

in your country, who carries out their function? Appraise their performance.

8. Find out the sources of capital of a specific private marketing enterprise in your country. On what terms can it obtain credit from a commercial bank and from the best alternative source of finance open to it?

9. After four years of record-keeping Kallu in Sierra Leone found that his sales had doubled, but his profits were halved. What were the reasons for this? What action would you recommend on the part of Kallu? The government? An aid agency with an open mandate for development assistance? What factors govern establishment and expansion of such enterprises in your country?

10. What is the attitude of your government towards indigenous private enterprise? Does it have a consistent policy for their development? What credit, training or other services does your government or other institutions provide for private marketing people?

4

Transnational enterprises

Pepper for Tabasco – Honduras

Some 150 years ago the McIlhenny Company obtained a variety of hot pepper seed from Mexico. They process peppers grown from this seed into a spicy sauce. Under the registered trademark 'Tabasco' it has gained worldwide market acceptance for seasoning foods and beverages.

For a long time, the peppers were grown in Louisiana and processed in the Company plant on Avery Island. More recently, Walter S. McIlhenny, the Company President, extended operations into the international market. By the late 1960s supply problems arose in the face of increasing demand. Timely harvesting is critical to the quality of the sauce, and labour was becoming scarce. Experiments with a mechanical harvester were not successful. The hot peppers have to be picked at the peak of their maturity and the pepper plant is a perennial bearing fruit nearly all the year round.

The Company looked for a suitable source of supply. After trials elsewhere it turned to Honduras. The Honduran Government welcomes foreign companies, knowing that both public resources and domestic venture capital were inadequate for creating more and better paying jobs in agriculture. Its experience with U.S. banana companies had been favourable.

Farmers in Honduras are accustomed to organize for a common purpose. When McIlhenny scouts made contact with a group of farmers in Yoro Province on the northern coast of Honduras in 1974 it did not take them long to form a hot pepper production co-operative (CAPHIL), and sign a contract. Unfortunately, they could not live up to their contractual obligations. Yields were low because of disease. The problem came to the attention of Steven Gross, a Jesuit priest who directed the Training Centre La Fragua in El Progreso, the principal town in the Province of Yoro. This Centre undertakes religious instruction, social education, and technical training. Following the devastation of the area by hurricane Fifi, it had received international relief funds to help finance agricultural projects. In 1976 Father Gross proposed a contract between the McIlhenny Company and the Training Centre La Fragua which would then arrange with the farmers to produce the peppers. The Company accepted this, advanced some funds for production loans, and a small plant to inspect, clean, trim, mash and salt peppers in wooden barrels was built on the property of La Fragua. In 1977, it operated with 55 groups of farmers supplying peppers. The Training Centre passed on to them cultivation technology and seeds supplied by McIlhenny.

Still technical problems continued. In addition to the wilt, farmers were not familiar with the crop, and some of their financing costs were high. Production costs were excessive. Half of the participating groups failed, defaulted on loans, and left the project in 1977 with a deficit of $17 500. In 1978 the number of farmer groups which applied for credit to La Fragua shrank to 13. A significant healthy change was made, however – interest-free credit was limited to the provision of inputs. In 1979 La Fragua had contracts with 51 groups which produced 300 tons of hot peppers. The farmers received 54 cents per kg. In 1980 the number of participating groups grew to 86. They produced 330 tons and received 56 cents per kg. La Fragua made a profit of $17 810. For 1981, contracts were signed with 90 groups representing 300 farmers to deliver peppers at prices ranging from 56 to 66 cents per kg. With over a hectare in hot peppers on average, the farmers were doing quite well in relation to the per capita annual income of about $500 in Honduras. On the rest of their land they grew corn, beans, rice, cocoa, tobacco, and raised cattle. The results were positive for the Company too. Exports rose from a value of $280 000 in 1976 to $753 000 in 1980 which reflected a combination of efforts by the farmers, the Training Centre and the Company. The farmers provided their land and labour while the Centre provided training, printed materials and demonstrations. The Company provided technology via an agronomist stationed at La Fragua, and visits by other company experts three or four times a year. Most of all, the Company provided an assured market at a fixed price.

With the project well under way, more facilities were established for primary processing. In addition to the plant in El Progreso, where La Fragua was paid $2.50 per barrel by the Company for storage, the Company financed a second plant in the Valley of Lean. The operating costs of the processing plants, including salaries of the 13 employees, were reimbursed by the Company which also pays 10 per cent of the cost of grinding to La Fragua in compensation of its expenses. In 1981 a third processing plant was under construction in El Negrito to reduce the transport costs.

Initially, rejection rates of 50 per cent were common. La Fragua technicians made sure that growers were present in the packing shed when their peppers went through the sorting and grading line. Growers watched the women on the line as they removed imperfect peppers so that they could see what standards were being applied. They could see that they were uniform and not capricious. This helped them enforce stricter standards on their own farms, so that they did not spend money sending inferior peppers to the packing shed.

In January of 1981 an advertisement appeared in the local newspaper. The Honduras Industrial Food Company offered to buy hot peppers at 73 cents per kg, green or ripe. The farmers found this opportunity hard to resist. Not only was the price 10 per cent higher than the top price paid by La Fragua/McIlhenny, there was the additional advantage of selling green peppers instead of waiting another 1½ months for them to mature. Deliveries to La Fragua processing plants dropped off, in spite of contractual agreements. McIlhenny was seriously considering withdrawing from the operation. To counter this, La Fragua made new contracts more legally binding. An information campaign was launched, with more frequent visits to farmers, to explain the benefits of participation in addition to price per kg.

Impact of the project

The Company obtained peppers at relatively low cost. It also achieved geographical diversification of supply, an important protection against ecological or disease problems. From the viewpoint of Honduras the project provided employment (12–20 people at the packing shed, as many as 300 farmers, jobs in transport, etc.). It produced a high added value per hectare, used semi-skilled and unskilled labour and relatively little capital; it diversified the agricultural exports of Honduras, and brought in a new technology which may lead to future developments. The project also had the advantage of being inherently suited to small-holder cultivation and participation. Because hot peppers are susceptible to disease, it is best to grow them in blocks of no more than half a hectare, separated by distances of 2-5 km. All available economies of scale can be captured with only a few hundred plants. Mechanical harvesting is not advantageous. A further advantage of hot pepper is that, once packed in salt for transport to Louisiana, it is no longer highly perishable.

After 1981 La Fragua ceased to be involved but McIlhenny Co. continued to contract for hot peppers from farmers trained by it.

Cassava exports from Thailand

Cassava has been grown in Thailand since 1850, primarily for human consumption. The decades, however, from 1957 saw a truly remarkable expansion. The area planted increased from 38 400 ha in 1957 to 960 000 ha in 1977. Output rose from 0.4 million tons in 1957 to 12.4 million in 1977. Exports at $528 million earned more than any other commodity.

The reason for this great expansion was the discovery of a market for cassava products as livestock feed in Europe and its development by transnational firms based there. Erich Funke, R. Schaller and the Overseas Barter Co., introduced Thai cassava meal to the European animal feed market in 1956. They saw the opening offered by the E.E.C. system of protective duties. These raised the price of locally grown feed grains far above the levels prevailing in exporting countries outside the E.E.C. Cassava was not subject to such duties because it was not considered a major source of feed supply for the Common Market. Also, there were no producers inside the Community with immediate interests to be protected.

Acceptance of cassava as a carbohydrate ingredient in animal feed mixes was also facilitated by the increasing sophistication of the compounding industry. If cassava was much cheaper than the carbohydrate element in barley then it would be used; the deficiency in protein could be made up by an additional input of soy meal.

The Germans then followed up with the invention of the cassava chipper and the establishment in Thailand of a German-made hammer mill to make cassava meal. Some of the early shipments to Europe suffered from mould growth. The great contribution of A. Toepfer was to overcome this by applying the process of pelleting flour with three per cent of molasses to cassava. In 1967–68 German investors put $1.0 million into the first pelletizing plant in Thailand.

Pre-eminent in organizing market flows have been Toepfer, Krohn and Co. and Peter Cremer. Krohn first thought of cassava as a livestock feed when in Thailand with the American forces. Cassava was being dried on some disused airfields and fed to pigs locally. He saw that they did well on it. Familiar with the European feed market from his family background he went back there to develop it. These firms traded in other feed ingredients for the European market. They had the necessary marketing expertise. They could easily mobilize finance so they were able to add cassava to their portfolio of commodities quickly and easily. In Bangkok, they are known as 'shippers'. They are in a position to charter whole vessels to transport cassava from Thailand to the European market. It is bulky and of low unit value, so low transport costs are essential.

This trade has also been very competitive. In the late 1960s it grew rapidly and the prospects looked bright so two of the biggest commodity transnationals, Cargill and Continental, came in. Only Cargill (Tradax) established itself. Continental closed down after two years with a rumoured loss of $3 million. In 1980–81 the Eurasian Corporation, formed by a group of Thai exporters, set up an office in Rotterdam and began selling directly to European feed compounders. It also lost $2.5 million during the first year but it kept going. By 1984 it had a 20 per cent share of the European market.

Market structure

The transnationals collaborate in using storage and loading facilities. They make joint representation to the Thai government regarding E.E.C. controls, but they compete in buying and selling. Each formulates his own market opinion and tries to make a profit from commodity trading.

Several have integrated horizontally or vertically. Krohn did both, by expanding its own pelleting capacity and also, in 1976, by forming a group of 16 or more Thai exporters into 'satellites' of Krohn. They buy pellets at a price set by Krohn each day and store the pellets in their own godown. The pellets are then exported either on Krohn's account or on their own account but to Krohn in the E.E.C. Krohn pays them a service charge. In this way pellet supply is secured and therefore market position.

Peter Cremer has concentrated on vertical integration, investing in a hard pellet plant in Thailand and in animal feed compounding factories in Europe. Alfred C. Toepfer has compound feed factories in Europe. Tradax (Cargill) invested over $20 million in port-handling facilities for pellets in Amsterdam. In 1981 it began building the largest hard pellet plant in Thailand. Table 4.1 shows the cassava pellets shipped from Thailand in 1978 by each of the transnationals, plus one local firm, together with the quantities provided by their suppliers. Some of these were wholly owned by, or associated with, one of the transnationals, others were independent. This balance had nothing permanent about it. In 1980, Krohn held on to its share. Toepfer's had increased to 21 per cent. Those of Peter Cremer and Tradax had fallen to 14 per cent each; 16 per cent was shared by other firms. Evidently favourable prices and margins had already brought forward the entry of new and effective competition. Erosion of the original transnational's market position was to continue with the entry of the Eurasian Corporation in 1981. By 1984 their share of the European market was down to 60 per cent. The number of local suppliers had expanded to 90 of which 50 to 60 could be considered really active.

Marketing operations

Growers sell directly to some 1500 cassava chipping plants which generally arrange for transport from the farm. This must take place immediately after harvest; fresh roots should not be held for more than four to six days. After sun-drying the chips are sold to pellet factories operated by firms such as those listed vertically in Table 4.1, known locally as exporters. Bulk handling is now widely used at the warehouse on to barges, and thence to terminal loading plants for ocean ships.

At the consumption end of the cassava marketing channel, buying begins with the formation of a production plan by animal feed compounders in the E.E.C. market. They use computerized linear programming methods to determine the least cost combination of ingredients to meet their nutritional formulae requirements based on whatever forward price information is available for each ingredient. To safeguard their supply position they buy forward for 2 to 12 months about 20 to 30 per cent of their needs. The balance is normally bought at the price of the day. Here, much depends on 'market opinion' what the various participants in the market – buyers and sellers – believe will be the market situation at a specified point in the future. They take into account prospective supply and demand, transport costs, foreign exchange and interest rates, government regulations, and any other economic, political, social, or physical factors which affect price. Each trader thus makes his offers on the basis of his 'market opinion'.

The transnationals generally buy forward from the Thai exporters in Deutsche marks at the port, with insurance paid and quality as

Table 4.1. Cassava market structure, 1978.

	Krohn & Co.	Tradax Ltd.	Peter Cremer	Alfred C. Toepfer	Trakulkam Feed	Total
			(Thousands of tons)			
Sui Heng Lee Co.	27	15	404	53		499
Sang Thai Bangpakong Co.	422					422
Charm & Sons Co.		417				417
Keng Seng Co.	35	248	46	69		398
Chaiyong (1970)	50	210	137			398
Soon Hue Seng Co.	106	60	30	140		336
Sahaphan Plant Products	116	90	25	88		319
P.H. Development Co.	184	71	18	35		308
Willing Trading Co.			298	6		304
Lo Chin Seng Co.	92			152		244
Peter Cremer			196			196
Poon Phol Co.	5		103	64		172
Krohn and Co. Bangkok	171					171
Thai Farmer Ltd., Part.	136					136
Thai Wah Co.	18	11	3	91		133
L.H. Co.	52	51		25		128
Sahathai Trading Co.	110					110
A.R.K.	101					101

Central Grain Co.	19	17	33	31		100
Thai Long Export–Import	96					96
Tai Long Ltd.	90					90
Bangkok Grain			64	21		85
Saeng Pet Import–Export	82					82
Union Eastern Product	18			63		81
Chai Charoen Exproduce	48			28		76
N.S.P. Thai Tapioca Co.	10			65		75
Lee Targ Seng				61		61
Thai Bamrung Thai	54					54
Song Kij Thai Trading Co.		20	16			36
Siam Tapioca and Produce		34				34
Thai General Tapioca	28					28
Trakulkam Feed Manufacturing					24	24
F.C.T. Brothers Co.		21				21
Thai Pellets Co.					19	19
Soon Hua Seng Produce Co.		17				17
Total	2070	1282	1373	992	43	5272
Percentage share	36	22	24	17	1	100

Source: Calculated from the sources of the Office of Commodity Standards.

delivered (D.Q.) at European ports. The transnational pays for lighter-age, stevedoring, and freight. The Thai exporters insure the cargo, and are responsible for quality which is examined at the European ports. After a sale contract is finalized, a letter of credit will be opened by the transnationals' European office to the Thai exporter. The exporter will then pledge this at the Bank of Thailand and obtain 90 per cent of its value subject to a special interest rate of only 7 per cent, with an obligation to export the specified shipment within 180 days. This low interest rate was established by the government to promote agricultural exports.

Any cargo not already sold forward to European buyers may be sold 'afloat'. This means that the owner starts to sell when the vessel leaves Thailand and may continue to sell until it reaches a European port. If he still has unsold cargo when the vessel arrives at European ports, then it will be sold on the spot market there. Selling afloat or spot in Europe is risky. Transnationals rarely store cargo in Europe; so the buyer usually has an advantage in this kind of sale.

The quality of a cargo is then checked by certified survey companies at the European port where it is unloaded. This is the condition of a 'delivered quality' contract. The inspected and certified quality and the actual weight are used to determine the discount or premium payable to each of the Thai exporters who supplied cargo. The shippers normally load one export cargo per hold (hatch) so each lot can be traced to a specific exporter.

Costs and margins

Average prices and costs for cassava pellets shipped to Rotterdam in 1982 are shown in Table 4.2. The price to the producer is for the unpeeled root equivalent, i.e. $2\frac{1}{2}$ tons. On this basis growers received about 37 per cent of the c.i.f. price in Europe. These figures do not permit separation out of the net margins taken by successive market participants, i.e. processor, Bangkok wholesaler/exporter, trans-national shipper, and distributor in Europe. The degree of integration must also be taken into account. Thus a firm which operated a chipping and pelleting plant, operated its own local wholesaling enterprise and also chartered the ship to Europe, would have a total net margin comprised of the profits on each of these activities.

Overall benefits

Cassava growers certainly did well during these years. Nearly one-third of the price they received was net margin after costs. Such profit opportunities lay behind the remarkable expansion of output. In addition to the potential market demand converted into a reality by the transnational and associated Thai intermediary and financing enterprises, the growers enjoyed production advantages such as a) easy and cheap propagation, b) relatively high yields, c) low costs of production, and d) good risk aversion.

Rapid growth also occurred in the sector concerned with the transport, handling, processing and warehousing of cassava. This now makes a major contribution to the economic activity of the country.

The transnationals made a unique contribution to agricultural development in Thailand and to its foreign exchange earnings. They made major investments, devised and brought in new technology, and saw and developed a market that was not recognized before. Changing market shares over time suggest that competition continues to prevail.

A brake on this progress was insistence by the E.E.C. in 1982 on the voluntary restriction of cassava imports. The Thai government was obliged to apply a quota system. Currently export quotas are allocated seven times a year taking into account the physical availability of supplies in godowns near Bangkok. Side effects from this procedure are higher storage costs and also sea freight charges because smaller vessels are chartered.

Table 4.2. *Prices and margins, cassava exports Thailand to Europe, 1982.* (Per ton of pellets and pellet equivalent)

	$
Price to producer (for 2½ tons of unpeeled roots)	37.25
Handling and transport of roots	3.70
Price at factory	40.95
Processing margin	18.52
Transport to Bangkok	7.78
Storage, handling, loading costs and margins, Bangkok	10.37
Price f.o.b. Bangkok	77.63
Export tax	1.71
Freight to Europe	20.37
Insurance	0.30
Price c.i.f. Rotterdam	100.01

Jamaica Broilers

Jamaica Broilers, was founded in 1958 by three men who were importing iced broilers into Jamaica. When weekly imports reached 100 cases of 27 kilos each, they decided to produce the chickens locally. From the outset, they adopted the 'contract farmer system'. Agreements were made with farmers who would be responsible for building a broiler house to company specifications, purchasing the equipment, and caring for the chickens during the eight to nine weeks required to bring them to marketable size. The Company, on its side, would supply day-old chicks, feed, medication and technical services at no cost to the farmers. The chickens, throughout their stay on the farm, would remain the property of the Company.

The Company put up its first chicken house with a capacity of 6 000 chickens. The contract system was initiated. The enterprise flourished as demand increased with the change over in the retail marketing system from the local grocery store to supermarket-type operations.

In the early stages this was very much an American-dependent venture. The contract system, the technical know-how, day-old chicks, ready-prepared feed, medicines, equipment, finance – all these were brought in from the U.S.A. Gradually it reduced its dependence.

Although at first it was convenient to import day-old chicks from Miami, this eventually became a major problem. The chicks were actually hatched in Georgia (some 1 500 km away), trucked to Miami and finally flown to Jamaica. All this stress produced a high mortality rate. When the volume reached approximately 12 000 day-old chicks per week, a decision was taken to set up a hatchery and import fertile eggs.

This decision, and another to build a modern broiler processing plant, marked the point at which the business exploded (see Table 4.3).

Table 4.3. *Broiler production in Jamaica.*

	Jamaica Broilers	Total national production	Jamaica Broilers market share
	(thousand tons)		(per cent)
1960	0.6	2.6	23
1965	1.7	4.3	40
1970	4.8	12.1	40
1975	8.3	27.5	30
1980	16.7	29.7	56

Since 1965, Jamaican broiler production has grown at around 15 per cent annually.

Feed mill

With this sort of growth, it became increasingly difficult for Jamaica Broilers to import sacked feed. Not only did it have to be ordered three months in advance, but shrinkage during transport to the farmers had also become a major problem.

In 1968, Jamaica Broilers and Central Soya, Inc., its feed supplier, entered into a joint venture to build a feed mill in Jamaica. Equity was divided 55 per cent for Central Soya and 45 per cent for Jamaica Broilers. Initially, the Company used Bank of Nova Scotia and Barclays Bank for working capital. The major investment in the feed mill, however, outpaced the ability of the local offices to handle Jamaica Broilers' financing. Accordingly, a five-year term loan from Barclays Overseas Development Corporation in London was obtained to finance the feed mill. Named Central Soya of Jamaica, Ltd the feed mill was completed in 1970. From that time until 1978, feed sales increased rapidly, not only to Jamaica Broilers but also to other livestock farmers throughout Jamaica.

The relationship between the partner companies was at times a difficult one. With the change of government in 1972, Central Soya became unwilling to make any major investments in the plant, as it opposed certain actions the government was taking. Finally, in 1978 with the assistance of a $1.5 million loan from Citibank N.A., Jamaica Broilers bought out Central Soya. The Company was renamed Master Blend Feeds Ltd, a wholly owned subsidiary of Jamaica Broilers. Now, with a more favourable political and economic climate in Jamaica, Central Soya would like to buy back in again.

Government controls

During 1978 to 1980, foreign exchange availability became a severe problem. Jamaica Broilers, with its near total dependence on imported supplies, was very vulnerable. Importing feed to produce broilers in Jamaica put value-added into the hands of growers and the national economy. In terms of food value alone, however, it was probably cheaper to import frozen chicken backs and necks available from the U.S.A. at very low prices. New feed ingredient formulae were found whereby the tonnage produced in 1980 exceeded that of 1979. Similarly, to make dollars for packaging go further, Jamaica Broilers began to use onion bags to hold frozen chickens at one-tenth the cost of the wire-bound freezer crates which had previously been imported from the U.S.A.

As early as 1969, broiler meat was brought under specific government price control. Since 1973, the number of broiler companies operating in Jamaica has gone down from six to two. This reduction has to a great extent been caused by rigid price controls. According to Managing Director Wildish, 'When price controls are fixed at the level of the producer of average efficiency, it is almost inevitable that you will eliminate those who are less efficient. If this process is repeated each time the price of chickens is increased, there is a real danger of creating a monopoly by driving too many producers out of the marketplace'.

Grower contract system

In mid-1981, Jamaica Broilers had 260 contract growers with an average of 14 000 birds on the farm. The growing cycle for all broilers averaged just under three months.

It is the responsibility of the contract growers to provide labour, utilities and management, and to construct all buildings to the specifications of Jamaica Broilers. The Company provides the feed, chickens and medications, and its advisory staff of two veterinarians, a poultry nutritionist and an eight-person field team.

The farmers receive three types of payments: rental, performance, and government payments. The rental payment is based on the cost of building and equipping the broiler house. There are five categories based on when the house was built and the materials used. The rental agreement guarantees a weekly payment to the contract grower whether or not birds are in the house.

The performance payment is based on the average live weight and feed conversion recorded for the flock. Paid when the birds are ready for slaughter, this payment reflects how efficiently the farmer used the raw materials of day-old chicks and feed supplied him.

The third type of payment is related to the pricing policies of the government. When the government increases the consumer price of frozen broilers, a certain percentage of this increase is returned to the contract grower. This payment can be up to $0.90 per bird.

Employee ownership

In 1973, Larry Udell, one of the three original partners, wished to capitalize on his initial investment. By this time, Jamaica Broilers was owned 70 per cent by Sydney Levy, one of the original Jamaican partners; 25 per cent by Udell, and 5 per cent by other local directors. The other original Jamaican partner had earlier sold his shares to Levy.

The Jamaica Broilers Employees Trust was set up to purchase Udell's shares with money lent by the Company. By 1977 it owned 25 per cent of the shares of the Company. To bring in contract farmers and owner/operators of the trucks that worked for Jamaica Broilers, the Levy family made available a further 30 per cent of the equity which would be offered to the contractors.

In April 1977, shares were offered to all employees and contractors as the Company continued to grow. Over 90 per cent of the people eligible participated in the share offer. Shares were paid for over a period of five years by deductions from earnings. If employees or contractors left the Company, the shares were to be sold back to the Trust. In 1979, the employees of Master Blends Feeds were offered shares in the parent Company, Jamaica Broilers, and all eligible employees participated in the purchase.

Jamaica Broilers currently employs 450 people, with an additional 260 contract growers and 36 truckers. The entire staff is Jamaican. Dr Wildish attributes the continuing success of the Company to the fact that so much of the profits has been re-invested, and to the employee ownership plan which has mitigated labour animosity toward management, sometimes a problem with Jamaica's extremely strong labour unions.

Fig. 4.1. 'The best-dressed chicken' brand sign of Jamaica Broilers.

The future

Jamaica Broilers feels that the domestic market for broilers has stabilized at 15 to 16 kilograms per capita per annum. It cannot compete in export markets as long as it has to import feed ingredients. For these reasons, the feed mill is expected to be the centre of the Company's growth as the total livestock industry in Jamaica expands. A two-year capital expansion programme costing $10 million is underway to increase the capacity of the feed mill from the current 112 000 tons per year to 170 000 tons with a fully computerized batch system. The processing plant is planned to remain at 7 200 birds per hour and the hatchery at 460 000 eggs per week.

Sales promotion has been very good. While few people in Jamaica would recognize the name 'Jamaica Broilers', everyone knows 'The Best Dressed Chicken' (Fig. 4.1) Jamaica Broilers' successful brand name. Legend has it that the phrase originated with one of the Company's founders who was an immaculate dresser. However, Dr Wildish considers its greatest success to be the development of the contract production and marketing system. He would like to extend it to fresh-water shrimp, fish and dairy cattle. Without hesitation he notes 'local ingredients are the key to our future'.

Cadbury in India

In British Commonwealth countries Cadbury has always had a high reputation. It is not only known for the wholesomeness and popular appeal of its products. The founding family, whose descendants still have a leading role in the Company, also carry a quality connotation. They belong to the community of Quakers whose members have had a formative influence on many great British industrial and banking enterprises. They had other values besides running a successful business. In 1879 George Cadbury moved his chocolate making out of grimy industrial Birmingham to a wooded site in the country. At Bournville he built a new factory and residential quarters for his staff surrounded by fields and trees. It became a model for 'garden cities' to follow. The Company now operates worldwide. In 1985 its turnover exceeded $2 800 million; its staff numbered more than 33 000.

Its Company in India began as a branch importing bulk chocolate products from the parent Company for re-packing and distribution. It took up local manufacture when India became an independent country, importing cocoa from West Africa – its traditional source of supply.

In 1978 new legislation limited foreign participation in companies operating in India to 40 per cent. I.B.M. and Coca Cola abandoned their operations in India. Cadbury decided to go along with the Government. Cadbury (India) Ltd became a public Company. The balance of its capital was put on to the Bombay stock exchange in hundred rupee shares, later subdivided into units of ten rupees (one dollar). In 1986 it had 23 000 Indian shareholders. All of its directors but one were Indian and all of its staff.

In face of press attacks on transnational enterprises its name was changed to Hindustan Cocoa Products Ltd. The Company continued, however, to use the brand name Cadbury – its most precious possession, and the parent Company continued to have a guiding influence on policy as a condition of allowing access to its brand. There is an agreement that each Company will share its information and experience with the other.

Financially, the parent Company is operating within a legislative framework set up by the Government of India which offers it very little direct incentive. It can take out of the country only the dividend declared on its shares in the Company – less 25 per cent withholding tax, but no royalties on brand names or processing technology. For the years 1985 and 1986 these foreign currency transfers averaged $270 000. The bulk of the net earnings of the Company are either ploughed back into the Company or taken by the Government of India in taxes.

Financial statistics for recent years are presented in Table 4.4. For the period 1978–83 retained earnings averaged 131 per cent of the amount paid out in dividends. Taxation paid by the Indian company took 61 per cent of its profits before income tax. The policy of re-investing in the Company a substantial portion of its net earnings does result, however, in a steady increase in the financial strength of the Company. A new plant with capacity to manufacture 2 200 tons of chocolates was set up at Induri near Pune. Earnings per share also increased fairly sharply from 1981 onwards.

Operations

The Indian company built up its sales steadily from $220 000 in 1948 to $48 million in 1986. Details for this year are set out in Table 4.5.

Cocoa beans are roasted, winnowed and ground into a mass. For chocolate this is mixed with milk and sugar to produce crumb. For cocoa powder the mass is pressed to bring down the fat content. The cocoa butter expelled is added to the crumb at a later stage to make

Table 4.4. *Financial statistics: Hindustan Cocoa Products Ltd.*

	$ millions		
	1978	1980	1982
Assets employed			
Fixed assets	4.96	8.30	6.54
Net current assets	7.02	8.83	6.24
Others	0.25	0.22	0.19
Total	12.23	17.35	12.97
Financed by			
Equity share capital	3.02	3.80	3.19
Reserves and surplus	4.10	5.20	4.46
Shareholders' funds	7.12	9.00	7.65
Loan funds	5.11	8.35	5.32
Total	12.23	17.35	12.97
Profits and appropriations			
Sales	27.40	24.43	40.48
Net earnings	2.16	1.33	4.76
Depreciation	0.33	0.59	0.53
Profit before tax	1.83	0.73	4.24
Taxation	1.20	0.08	2.69
Profit after tax	0.63	0.66	1.54
Dividends	0.40	0.23	0.51
Retained earnings	0.16	0.42	0.97
Ratios			
		Dollars	
Earnings per share	0.26	0.22	0.52
Net worth per share	2.87	3.00	2.55
		Per cent	
Earnings on shareholders' funds	9	7	20

chocolate. The secret of success in chocolate making is to have sales of chocolate and of cocoa powder in balance. Cocoa powder is a principal ingredient of Bournvita which, mixed with hot water, made an attractive drink of significant food value. Other ingredients were malt, milk and eggs. Malt was produced from barley processed at a plant at Induri. A dairy farm was set up there to provide the milk. It still carries 800 head of cattle predominantly progeny of a Jersey/local breed cross. A poultry farm was also set up to produce eggs. Later, this ingredient was dropped because of the strong vegetarian feelings of many Indian consumers.

Table 4.5. *Product sales: Hindustan Cocoa Products Ltd, 1982.*

	Tons	$ millions
Malted foods	4357	14.1
Cocoa powder	20	0.1
Drinking chocolate	262	1.0
Chocolates	3146	19.8
Cocoa butter substitutes	1571	4.5
By products	276	0.2
Total		39.7

Through the 1960s and 1970s output was restricted by an import quota of 1 000 tons of beans. Imports were also subject to a duty which reached 120 per cent in 1976–77. This meant that the chocolate products marketed were priced beyond the means of the majority of Indian consumers. It was this straitjacket on their operations that led Cadbury to take the initiative in promoting cocoa production in India. When the breakthrough came in the 1980s, and following a period of crisis, the purchase price was brought down to world market levels, output shifted on to a raw material base of 2 000 to 3 000 tons annually.

Promotion of cocoa production

The great achievement of Cadbury in India from the point of view of its farmers and its rural economy was the development of domestic cocoa production. Through the initiative of the then Managing Director, the Indian Government invited D. H. Urquhart in 1958 to investigate the prospects of cocoa growing in India. He recommended the Forastero variety in place of the Criollo variety which was already grown in India on a small scale, but has only a limited commercial market for flavouring purposes. However, in 1962 the Government of India decided that Criollo should be propagated in Southern India, the only part of the country where rainfall is sufficiently distributed over the year to favour cocoa growing. Nevertheless the Company persisted. It appointed an Indian cocoa advisor and leased land for a research and demonstration farm. It paid for the import of Forastero seed from Malaysia. In 1972 this was stopped by the Government of Malaysia; work continued with indigenous planting material.

Initially Cadbury gave cocoa seedlings free to potential growers together with a commitment to buy their eventual crop. Most of it is grown on smallholdings of two to four hectares intercropped with coconuts and/or areca nuts (source of a stimulant used in betel nut chewing). A network of buying depots, collection centres and drying units was established. Purchases of wet or dried beans were made for cash. Cadbury is now buying 2 000–3 000 tons annually from some 5 000 to 6 000 farmers. At the price paid in the 1970s of $3 per kg, a yield of 500 kg per hectare gave a gross return to the farmer of $1 500. Assuming he already had the land, with labour costs estimated at $90 per hectare and outlay on fertilizer $55 this was a very attractive return.

The price was one that could not be maintained. It reflected the high world market prices for cocoa in the 1970s, doubled for imports into India by the duty imposed. Cadbury felt obliged to continue paying that price after world market prices declined, until its factory was halted by strike action in 1980. Surpluses of beans built up for which the only outlet was export. Eventually the government intervened and halved the price. Now Cadbury pays a price that is equivalent to the world market after allowing for a 10 per cent discount on Indian beans because of their prevailingly small size. Production continues to increase.

Because of the small size of the producer units, Cadbury buys most beans wet and does its own fermentation and drying. Its field staff consists of the Chief Cocoa Adviser with six assistants located in the cocoa growing areas. Their original focus was on extension, but, with a buying clerk and labourers, they also organize purchase from the growers and local processing. Driers are used, though some cocoa is moved for drying to areas that are naturally sunny. The dried beans are then trucked to the factory near Bombay. The dried weight recovery from wet beans is about 30–35 per cent.

Barley for malt is bought from four or five wholesalers. Milk is bought regularly from farmers in the area near the factory to complement the supply coming from the original directly operated farm.

Distribution

Distribution of chocolate products proceeds through local wholesalers to retailers. Deliveries to wholesalers are made by carriage and freight agents using insulated vans cooled by blocks of ice. Illustrative prices are presented in Table 4.6.

All of the staff of the company are now Indian. Yoginder Pal the managing director and chief executive came to the company from

Unilever in India. The director of personnel Aroon Joshi, brought in after the strike in 1980, worked previously for Fiat and Metal Box.

The sales and marketing director, Aadhi Lakshmanam has been with the Indian Company since 1949. A radical change for him was the shift to cash transactions with wholesalers. The traditional practice of accepting a deposit and then releasing stocks on credit was stopped two years ago to improve the company's cash flow. In India, Cadbury has a very strong market position with 85 per cent of chocolate product sales. Its competitors are essentially local – led by the Amul Dairy Co-operative Federation which has received technical assistance from Nestlé.

The strategic link with the parent company is Andrew Harvie-Clark who is chairman of the Indian Company and, at the same time, in charge of Cadbury's Asian operations. He travels frequently between his bases. Staff training arrangements include the stationing of an Indian research specialist in the Cadbury laboratories in the UK and a stay by the factory manager at Bournville.

Overall appraisal

Cadbury in India has made available to the people of that country food products and chocolate confectionery in widespread demand. It has maintained a high standard of quality. Primarily because of this it has a dominant share of the market.

For much of the period since its inception it has faced an adverse government climate. Its products were not regarded as a high priority. Like tobacco they were seen as a legitimate subject for high taxation. Its

Table 4.6. *Product prices to wholesalers and retailers: Hindustan Cocoa Products Ltd, 1984.*

	Price to wholesaler $ per unit	Retail price
Milk chocolate 80 g	0.47	0.64
Fruit and nut bar 40 g	0.27	0.36
Eclairs 100 g	0.44	0.58
Bournvita 500 g	1.57	2.0
Cocoa 200 g	1.0	1.32

import requirements were also viewed unsympathetically. The spin-off from this was the development of a durable, profitable domestic cocoa production. This has very largely eliminated the need to use foreign exchange on imports. Cocoa is commonly intercropped between coconuts and areca nuts. It does not take land away from food production. Most of it is grown by small farmers to whose incomes it adds substantially. Here also, the Company had to persist with the Forastero variety against a negative government position. Ultimately it was the government that reaped most of the financial returns through the tax revenues Cadbury provided. Because of this high taxation the direct income to the parent Company is paltry. Its primary considera-tion is to keep its name before the eyes of consumers in one of the world's largest and most populous countries. A favourable image there, with participation in a sound and expanding enterprise, keeps it in good position to take advantage of new opportunities that might open up in the future.

Unilever-Is copartnership – Turkey

Unilever, with its headquarters in London and Rotterdam, is one of the largest of the multinationals in food and agricultural marketing. This is both in terms of business handled and the number of countries where it operates.

In Turkey its main activity has been the preparation and marketing of margarine and cooking fats based on domestically produced oilseeds when available. This has been developed through Unilever-Is Ticaret – a Company in which 35 per cent of the capital is in Turkish hands and which is now nearly entirely managed by Turkish nationals. In 1953 it marketed 770 tons of margarine, 5 000 tons of primarily vegetable oil-based ghee and 250 tons of other fats. These quantities were built up through the 1960s to 16 500 tons of margarine sold under the brand name Sana and 46 000 tons of ghee marketed under the name Vita. At the end of 1983, sales of Sana margarine at 103 000 tons were the largest in the world for one brand. Sales of Vita had declined to 3 000 tons with the shift in consumer interest to a higher value product. An easier spreading margarine Rama sold in a plastic carton had been introduced and now sold 4 000 tons. Including other fats, total sales amounted to 120 000 tons. The main raw materials for these products in Turkey are cottonseed and sunflower seed oil. Imported soya oil is also used when domestic oilseed supplies are insufficient. However, the demand for the sunflower seed oil, in particular, created by the Unilever-Is enterprise has been largely instrumental in stimulating domestic seed production expansion from 50 000 tons to 700 000 tons.

Effective Unilever interest in Turkey began with a 1949 mission by P. Van den Bergh and J. J. Clerk of the Dutch side of the business that is primarily concerned with margarine. Their report noted that there might be an important market for 'a cheap and wholesome substitute of butter and clarified butter as both these articles are outside the reach of the large majority of the population'. They estimated this market at about 1 500 tons of margarine and 1 000 tons of ghee substitute, with the possibility of considerable further expansion after three or four years. 'We have a reasonable chance of having the market to ourselves for some years, or at least until we have been able to build up a strong position through the creation of a goodwill for our advertised proprietary brands.' A local firm had been interested. But it still had not learned how to make margarine, though it was said to have spent a great deal on new equipment. Most important the Government was very eager to see a margarine and vegetable ghee industry established in view of the shortage of butter and ghee. The raw material would be sunflower oil whose production in Thrace was increasing rapidly. The outcome was a partnership between Unilever and the Is-Bank in Turkey which was to be central to the success of the business. The bank suggested that Unilever should provide $1.4 million and the Is-Bank $250 000 plus $1 050 000 as working capital. The Is-Bank could be expected to provide political as well as financial support.

The prospective benefits to the country were set out as:
(a) By using almost entirely indigenous raw materials, it will create a new outlet for agricultural products.
(b) By making available products of a high nutritive value, it will increase the general standard of health in this country.
(c) By a policy of reasonable selling prices, it will assist the Government in their aim of reducing the cost of living.
(d) It will, once the products have been accepted by a large part of the population, including the armed forces, free considerable quantities of oils, such as olive oil, for export.
(e) It can, particularly if exports could be arranged under Turkey's trade agreements with other countries, play a certain part in the Government's export drive.

The Government welcomed the proposal.

A mixed team of foreigners and Turks was appointed to manage the

Fig. 4.2.

enterprise. At this stage the full value of a multinational company became obvious in the flow of experts who arrived in Istanbul to help the nascent company. An expert on marketing margarine, spent 18 days in the autumn of 1951 investigating distribution problems. His visit overlapped with one who came to discuss the purchase of raw materials. Another specialist analysed promotional possibilities. His conclusion was that the budget should be divided in the ratio: press 67 per cent, radio 14 per cent, shop demonstrations 5 per cent, exhibitions 9 per cent, display material 4 per cent and printed matter 1 per cent. The emphasis on the press was a tribute to the increase in literacy. The radio budget was relatively small partly because Turks without a main electricity supply saved their batteries by turning them off during advertisements. Figure 4.2 presents an early advertisement of Vita. Figure 4.3 is one for Sana in the 1980s. By this time television had

Fig. 4.3.

become a much more important vehicle than the press for Sana advertisement. The focus was on mother-care for children.

The obstacles that had to be overcome by Unilever-Is during its early years were considerable.

It had to surmount:
(a) legal and technical difficulties relating to the nature of the product;
(b) problems in obtaining essential supplies, and
(c) it had to create a market for its product.

In 1950 a draft law intended to define margarine was remitted for consideration to a committee of three senior Turkish scientists. Their conclusion, that margarine must be made by 'shaking fat with cream', surprised and disturbed Unilever men because the standard European method of making margarine was with water, with milk used as a flavouring. They had considerable difficulty in persuading the committee of the better preservative qualities of margarine made by the European method. The Turkish scientists also accepted the use of a preservative and the addition of vitamins, together with proposals to reduce the minimum fat content from 86 per cent to 80 per cent and to fix 36 °C as a maximum melting point of each fat used rather than restrict it, as originally proposed, to a lower than human body temperature. The Chairman also promised to give favourable consideration to the use of carotene, butyric acid and ethyl butyrate as colouring and flavouring matter.

At one stage margarine was classed as a manufactured product on the grounds that it contained added vitamins, flavouring and colouring, and this made it liable to a 15 per cent tax. Eventually an exemption was obtained. Finally the company had to face a technical difficulty that resulted from using sunflower oil. At the start the sunflower oil when hydrogenated became brittle and wet and this caused the margarine to fall to pieces. This also was overcome.

In 1958 it was reported that stocks of packing material and chemicals were exhausted at the factory, but sufficient for 4–5 weeks was waiting at the Customs. Release would take place as soon as a licence was signed by five ministers; those of Finance, Commerce, Industry, Customs and the Minister of State, plus a President of the Bank of Turkey. The factory was idle for about two weeks. J. J. Clerk recalls that he often spent whole days from 5 a.m. sitting near the lift in the Hilton Hotel, Istanbul, or some other place, hoping to catch a Minister and get him to sign a vital document. The reason for these delays lay in the chronic balance of payments problems experienced by Turkey during these years and in its bureaucracy.

One of the first successes in developing a market for the Unilever-Is products was an order of 300 tons of vegetable ghee for the army and 20 tons of table margarine for army hospitals, prisons, etc. Van den Bergh exulted: 'Our brand name on all packets and tins will get known through the soldiers to every hamlet in Turkey.' He was right. The Vita tins, which were used for many purposes including (after being flattened) roofing houses, were the means by which publicity spread in many parts of Turkey which could not have been penetrated by normal advertising for a very long time. Ghee was a product already in general use; and Vita, as a much cheaper substitute, was almost certain to sell well as soon as the public became aware of its quality and price.

The original estimates of market resistance proved unduly pessimistic. Sales expanded at an astonishing rate. This was certainly not due to intensive promotion since advertising expenditure was exceptionally low. Probably the two main reasons were rising demand and the relatively low price of Unilever products. Demand for edible fats rose as the population increased from 21 million in 1950 to 31 million in 1965. Production of butter and natural ghee could not keep pace with this surge, partly because cattle were being replaced by tractors and pasture by arable. This created an expanding market for good-quality substitutes in margarine and vegetable ghee. A closely related factor was that prices of the animal products rose as demand out-ran supply. Unilever-Is, by contrast, was able to keep the price of its margarine and vegetable ghee well below that of butter and natural ghee because the cost of its raw materials did not rise as fast. Thus in January 1962 fresh salted butter of the Trabzon type cost $1.44 a kilo as compared with Sana margarine at 67 cents a kilo. Olive oil cost 89 cents a kilo while Unilever-Is sold Vita at 62 to 64 cents per kilo. These price advantages were very important in Turkey, where the average income was then estimated at $180 per annum. Low costs also helped Unilever-Is during periods of government price controls. Prices based on the production costs of its Turkish competitors still allowed Unilever-Is to make a profit.

Though living dangerously, Unilever lived well in Turkey. The financial results are summarized in Table 4.7. Between 1951 and 1965 issued share capital rose from $1.79 million to $4.62 million, of which 80 per cent belonged to Unilever, the rest to the Is-Bank. In the same period Unilever transferred $5.45 million in dividends and the Is-Bank received $1.36 million. No payments were made to Unilever in the form of service fees or royalties. The dividends paid by Unilever-Is were, however, only about a third of its total pre-tax trading profits. They were kept at this level partly because the business needed funds for expansion, partly because the Government was happy to allow that proportion of profits to be transferred. Of the trading profit for the decade 1956–65 – a total of $21 million – 8.6 million went in taxation, $5.8 million were retained in the business, $1.32 million went to the Is-Bank and $5.27 million was transferred as dividends to Rotterdam. Thus only about a quarter of the total profits made by the business was withdrawn from the Turkish economy; the Turkish government benefited more than Unilever from its activities.

One of the factors to which Unilever-Is attributes its success in Turkey is its efficient purchasing of its raw materials. These were primarily sunflower seed oil and cotton seed oil. The latter came as a byproduct of the cotton industry in southern and western Turkey. Unilever-Is bought the oil from the crushers. Sunflowers were grown specifically for the oilseed market primarily in Thrace. Again, Unilever-

Table 4.7. *Summary trading figures, Unilever-Is, 1956, 1965, 1983, 1984.*

	1956 $ millions	1965	1983	1984
Net sales	27.22	41.61	100.28	126.50
Gross profit	4.29	6.78	24.93	29.61
Costs				
Advertising	0.35	0.98	1.04	0.88
Marketing	0.63	1.17	0.41	1.29
Factory and general	0.70	1.62	6.24	8.35
Head office, research,				
pensions	0.20	0.61	2.00	2.71
Total costs	1.88	4.38	9.69	13.23
Profit before tax	2.31	2.40	15.24	16.28
Profit after tax	1.78	1.54	9.09	8.99
Gross capital employed	9.07	16.59	31.42	40.04
Costs as per cent of sales	*Per cent*			
Advertising	1.3	2.3	1.0	0.7
Marketing	2.3	2.3	0.4	1.0
Factory and general	3.0	4.0	6.2	6.6

Is bought mainly from local crushers, also seed from the Thrace oilseed growers' co-operative which was then crushed at a mill on contract. Most of the time the Government maintained a guaranteed minimum price for sunflower seed paid to farmers on delivery to the co-operative. However, farmers seeking credit might commit their crop by contract to a crushing mill at a discount from the Government price in return for the advance. Similarly Unilever-Is advanced funds to two cottonseed and two sunflower seed crushers on the understanding that it would receive all their production when ready. Thus Unilever was not so much a price setter for farmers as a consistent, sure outlet for the semi-processed product. Other factors influencing the supply were the relative advantages of growing wheat and sugar beet for which minimum prices were also fixed by the Government. Unilever-Is had, however, a continuing influence on quality. It maintained as an incentive to its suppliers a consistent set of deductions if oil failed to meet established standards of colour and freedom from fatty acid.

Concerned always over its raw material supplies, Unilever-Is has helped ensure availability of suitable seed. When in 1962 Turkey's sunflower crop was ruined by orobanche disease, the Company supplied Turkish farmers with an orobanche-resistant strain of sunflower seed with a higher oil yield which they obtained from Russia via France. Subsequently it funded research at Yalova University to develop improved seed. Nevertheless yields in Thrace, the most favourable growing area in Turkey, have still been only half those obtained just over the border. A direct initiative to overcome this constraint has been the establishment of a new Company with the Thrace oilseed growers' co-operative and an American seed breeder the Interstate Seed Co., to produce hybrid sunflower seed in Turkey. This Company imported 80 tons for planting and a small amount for multiplication in 1984. In expectation of higher yields and to meet unsatisfied demand that is expected to permit total sales of 150 000 tons, Unilever-Is has established a new plant at Corlu in the Thracian production area. The 1984 financial accounts show an increase in capital (to be supplied proportionately by Unilever and its Turkish partner) to cover the cost of this new investment.

Unilever-Is never established direct purchasing links with farmers for its raw material because of the high seasonal concentration of farmers' sales. Everything had to be bought in a three-month period for which a team of buyers would have to be maintained and cash and storage provided. At the distribution end, however, a steady flow was needed to meet consumer requirements throughout the year and close control of quality at the retail level was essential.

Unilever-Is maintained its own distribution network that delivered 75 per cent of its sales directly to the retailers. The frame for this was five wholesale depots and 11 sub-depots. Salesmen go out from these to take orders. Vans follow them the day after with deliveries. At one time it used 1 600 wholesalers. This number is now down to 700. The company prefers the direct retailer contact for its potential influence on the price charged, the sales display used, and, above all, quality. Stocks unsold after two months are taken back and remelted. Experience also shows that this system of distribution can be cheaper. The conventional mark-up has been 5 per cent for the wholesaler, 10 per cent for the retailer. Costs at the direct distribution depots have often stayed within the two to three per cent range.

Unilever takes pride in protecting its brands against variability in quality. This has been one of its greatest attributes in competitive marketing. Before the use of refrigeration became general in Turkey this called for considerable ingenuity. Unilever-Is pioneered the use of evaporative cooling based on local materials. Trays of margarine in transport were covered first with clean waterproof covers then with straw mats that had been soaked with water. If they dried out during a journey, the driver had to find water to keep them moist. Drivers were instructed to park vehicles under shade wherever feasible.

The staff of Unilever-Is is strongly Turkish. From the start it replaced Europeans by Turks as soon as possible. By June 1958, five years after the start of production, there were only ten expatriates out of 534 employees: by 1976 only the Chairman of Unilever-Is was an expatriate. In 1984 he was also a Turk – a man who started there as an assistant accountant in 1948 and acquired experience with Unilever in the U.K. and other countries. The Company, of course, offered a career open to local talent: an able Turk could rise to become a director. This was seldom possible for any but the family in the typical Turkish family business. A man who will head the new seed company in which Unilever-Is has a share studied chemistry at university in Turkey. He then obtained an advanced degree in fat technology in Germany. It was his professor there who recommended him to approach Unilever-Is for a career post.

To the question 'What factors were most instrumental in the success of the enterprise?' one of the Turkish managers gave the following response.

1. With its resources and established reputation Unilever-Is could attract and hold the most qualified staff.
2. Access to accumulated know-how, experience and research results through its parent company gave it a great advantage over domestic competition.
3. Specialists in different disciplines working together as a team resulted in very good management.

A further consideration has been consistently low overhead costs. Unilever-Is operated a single, compact, mechanized plant, kept down its headquarters staff, marketed a few products for which there was growing demand and relatively little competition. So its advertising and other marketing expenditures were low. With low costs and expanding sales it could generate high profits from low margins.

Mumias Sugar, Kenya – Booker McConnell

The partnership of Booker McConnell with the Government of Kenya in the Mumias Sugar Company (M.S.C.) has been described as one of the most imaginative and successful examples of transnational private participation in agricultural development. It provided productive employment for an area of the country previously neglected, was highly labour-intensive in its operations, made up a deficit in supplies of a major food that would otherwise have had to be imported, and paid over to the government in taxes a substantial part of its gross income. Even Susan George in her critique of the multinationals *How the other half dies* (Penguin Books 1976) had only good words for Mumias Sugar.

Origins of Mumias Sugar

Traditionally Kenya has imported much of its sugar. Consumption was growing by more than 7 per cent per year when, in 1965, the Ministry of Agriculture prepared a sugar project for the Nzoia river valley. The climate was suitable; also the region though quite densely populated was very underdeveloped.

Booker Agriculture International (B.A.I.) had been set up by the sugar processing and marketing firm Booker McConnell following the nationalization of its operations in Guyana. It had a subsidiary Fletcher and Stewart that built sugar processing plants and machinery.

Following an approach by the Government of Kenya, Booker Agriculture International, drew up a project plan for a 3 300 ha nucleus estate and an outgrower component. The Mumias Sugar Company was formed in June 1971. The factory began operations on July 1, reaching its designed capacity by September the same year. Between 1977 and 1979, factory capacity was more than doubled, from 125 to 300 tons of cane per hour. By 1979 Kenya achieved self-sufficiency in sugar with Mumias providing 45 per cent of its supply.

Capital structure and operation

Mumias Sugar Company was established as a commercial corporation with 3 500 000 authorized shares of 20 Kenyan shillings each. Booker Agriculture International took 5 per cent at the insistence of the Kenya Government to ensure that it had a financial interest at stake. Details of the shareholdings and loan capital in 1974 and 1981 are provided in Table 4.8.

In addition to carrying out a pilot project and performing a major design role, B.A.I. had a managing agency agreement with the Mumias Sugar Company. This, along with the contract to supply the sugar factory, was Booker's main reasons for going into the project. B.A.I.'s remuneration had three elements:

1) A relatively small fixed fee to cover the general manager's salary and B.A.I.'s relevant overheads. Other B.A.I. staff were seconded to the project at cost.
2) A commission on net M.S.C. revenues (i.e. sales of sugar to the Kenya National Trading Corporation less Government excise duty). Contrary to what might be expected in some quarters, this commis-

Table 4.8. *Capital structure: Mumias Sugar Company.*

	End 1974	End 1981
Authorized shares of K.sh. 20/–	3 500 000	12 000 000
Shares issued and fully paid	2 900 000	8 500 000
	$	$
Kenya Government	3 800 000	8 421 000
Commonwealth Development Corp.	661 000	2 044 000
Kenya Commercial Finance Co.	498 000	595 000
East African Development Bank	276 000	314 000
Booker McConnell Ltd	276 000	526 000
Total share capital	5 511 000	11 900 000
Loans and advances	4 560 000	21 284 000
Profit and loss account	793 000	(108 000)
Total capital employed	10 864 000	33 076 000

Conversion rates 1974 K 20 sh = $1.90: 1981 K 20 sh = $1.40.

sion rate increased with the volume of output. This was done to give B.A.I. a direct incentive to maximize output. Commission rates were first set as follows:

Annual output	Per cent of net revenues
less than 45 000 tons	0.0
45 000–50 000 tons	0.5
50 000–55 000 tons	1.5
55 000–60 000 tons	3.0
60 000–65 000 tons	4.0
more than 65 000 tons	5.0

3) $2\frac{1}{2}$ per cent of net profits of the Mumias Sugar Company. This was done to give B.A.I. an incentive to operate the project as efficiently as possible.

With the subsequent substantial expansion in production capacity these terms were revised later and made subject to an overall maximum.

The outgrowers' farms were small. It was considered socially desirable to use no more than 50 per cent of their area for sugar cane so that food supplies could be maintained. The Company grouped together individuals'cane plots to make stands of cane no smaller than 6 hectares. This enabled machines to be used economically for deep ploughing and other heavy land preparation work. Cane transport costs could also be kept down in this way.

Disease-free planting material and fertilizer were supplied to each grower on credit and each farmer was then responsible for his own planting, weeding, and fertilizing. However, Company representatives visited the plots regularly to advise farmers and harvesting was carried out by the Company. This control was essential to maintain the necessary regular volume of cane flow to the factory. All cane, both on the nucleus estate and on outgrowers' farms was cut by hand which provided more employment than mechanized methods.

This relationship between the sugar company and the growers was covered by a contract and at the end of 1981 17 474 farmers had signed contracts with the sugar company. The outgrowers' scheme was a clear success. The difficulties came up on the marketing side.

Pricing risk

The participants in the project knew there were risks on the production side, but the market seemed clear: Kenya was in deficit for sugar; its expanding population would take up whatever could be produced.

M.S.C. was to sell output to the Kenya National Trading Corporation at a price fixed by Government. The Corporation was obliged to buy all of M.S.C.'s production. Nevertheless, the problems came up on the market side. On occasion the Corporation found itself unable to take delivery of sugar already manufactured. M.S.C.'s own operations had then to be shut down temporarily for lack of finished sugar storage space.

Most serious for the Company was the Government's unwillingness to raise the price paid for sugar in spite of inflation. Between May 1977 and April 1981, despite repeated requests from M.S.C., the Government kept prices constant. The Company was forced to subsidize cane farmers' operations. The cane and sugar price increases approved in 1981 were 12.8 per cent and 9.8 per cent respectively, while over the same period the general rate of inflation was 48 per cent. The impact of this on M.S.C. can be seen in Table 4.9. Net profits before tax dropped from over $5 000 000 in 1977 to a loss of $108 000 in 1981. Yet production had been raised by 90 per cent to 167 400 tons. In his 1980 and 1981

Table 4.9. *Summary operating results: Mumias Sugar Company 1977, 1979, 1981.*

	1977	1979	1981
	$ thousands		
Gross turnover	29 780	40 848	48 847
Excise tax	7 936	10 490	11 831
Net turnover	21 844	30 358	37 016
Payment to growers	6 152	10 042	14 197
Profit (loss) before taxes	5 738	276	(108)
Income tax	2 527	(467)	–
Profit (loss) after tax	3 211	743	(108)
Equity	17 547	17 334	14 069
Profit after tax as percentage of equity	18%	4%	–
Dividends	2 430	1 620	–
Dividends as percentage of equity	14%	9%	–
Tax revenue to government	10 463	10 490	11 831
Nucleus estate cane (hectares)	3 300	3 300	3 300
Outgrowers cane (hectares)	9 900	15 400	24 300
Cane crushed (tons)	697 000	975 000	1 566 000
Sugar produced (tons)	81 275	109 800	167 400
Annual increase in production	28%	19%	2%
Registered farmers	9 372	13 113	17 474
Permanent employees	3 521	4 108	4 936
Seasonal employees	89	5 282	9 218

annual reports, the Chairman, Professor George Saitoti, summed up the situation as follows:

(1980) The company's cash resources have suffered over the past three years from diminished profits due to the lack of a price increase and the need to subsidize the farmers. In view of the present liquidity position, the directors have recommended no dividends for 1980. It is deeply regretted that, due to circumstances outside management control, shareholders will be denied the reward of their successful investment.

(1981) The Company's cash position continues to be highly unsatisfactory and survival was possible during the year only by taking excessive credit on the payment of the sugar excise.

Any private Company's operations in any country are exposed to the unfavourable effects of changes in policy or regulations by the Government. B.A.I. had minimized these risks by entering into a partnership with the Kenyan Government to pursue common economic and developmental goals in such a way that both Company and Government were rewarded if objectives were achieved. Unfortunately, the pressure of outside events, particularly increases in the cost of imported oil, resulted in the Government seeking every possible means to keep down price inflation, especially in basic food items like sugar. While the Government as a major shareholder had to forego dividends along with the others, and received no tax on Comapny profits because there were none, it continued to collect the excise duty. So, even after allowing for devaluation of the Kenya shilling, it took out more in dollar terms than in previous years.

Benefits from the Mumias project

The Company was profitable from the first six months of operation. This was due to higher cane yields, higher factory capacity utilization, and lower costs than had been forecast. The social cost–benefit analysis has also been very favourable, mainly because of the boom in world sugar prices during the early years of the project. By the end of 1974, a mere 18 months after production began, the project, based on the opportunity value of sugar at world price levels, had completely paid for itself. Sugar prices have fallen since. However, B.A.I. management estimate that Mumias has the lowest production costs of any sugar project in East or West Africa.

Almost 5 000 permanent and more than 9 000 seasonal jobs have been created, and almost 17 500 farmers are now receiving cash incomes who were not before.

Company employees have benefited from organized training at the apprentice, technician and management levels. By 1978, out of 102 management employees, only 17 were expatriates. In accordance with the original agreement, B.A.I. was to withdraw at the end of its fixed term management contract and Kenyans would take over complete managerial responsibility.

It is said that the project furthered inequality in land holding and incomes among the farmers concerned, led to relative neglect of traditional food crops and to social deficiencies deriving from rapid access to wealth. These criticisms cannot, however, be pressed very far. They are based on hindsight on a project that succeeded beyond all expectations.

For Booker McConnell the payment by results contract proved very favourable. Its management fees averaged a $1.7 million over the years 1980–82. This was about 3 per cent of gross turnover – not high, however, considering the initial management risks, the success achieved and the part played by its management in achieving it.

It was also a step to other such contracts in Somalia and Papua New Guinea, and eventual takeover of a majority share in the International Basic Economy Corporation (I.B.E.C.) of New York.

Attributes, advantages and support needs

International private marketing enterprises have played a major role in the whole process of development of many countries in Africa, Asia and Latin America. They had a market in the temperate countries for the products of the tropics. They provided transport facilities and other marketing infrastructure, undertook vital research and promoted production of the goods they sought. It is largely due to these enterprises that the export marketing of tea, coffee, sugar, bananas and rubber has been relatively well organized. They were able to mobilize capital, technicians and managers from the places where they were most plentiful. Because of their direct contact with buyers in the countries where their produce was consumed, they were acutely conscious of competitive quality standards and techniques of presentation to consumers.

Attributes

The characteristic positive attribute of the transnational is that it has at its disposal resources of technology, management, capital and market knowledge that are either not available, or not available at the same level, in a developing country.

Finance The transnationals interested in supplies of cassava chips from Thailand provided *capital* for the establishment of processing plants and port-handling facilities. Their letters of credit drawn on European banks were the basis for the financing of growers, assembly, processing and loading on to the ship. McIlhenny financed the assembly, processing and export of pepper from Honduras, plus credit for production inputs.

The *technology* of the transnationals has the great quality of being already applied to a particular commercial operation. Tested by experience it has very high commercial value. In reality, technology does not come in convenient self-contained units that can be shipped ready for use from one part of the world to another: technology is transferred through people.

The technology possessed by firms like Booker McConnell and Unilever is derived from a continuum of activities involving, not only scientific and production skills, but also marketing operations, procedures for quality and stock control, purchasing, accounting, distribution and personnel management. Central Soya supported Jamaica Broilers with both technology and capital.

Firms with operations in both developing and developed countries where they face strong competition bring continuing *access to new technical developments*. A common misconception is to see the transfer of a technology package as a once-for-all process. If this were true, developing countries would very soon be saddled with outdated processes and techniques. When Unilever implants technology as part of a business investment, the recipient can benefit from continuing access to the latest developments through the ideas, advice, information, people and expertise which flow regularly between the head office 'clearing house' and the operating companies. This system ensures that the various national companies get what they need.

Getting it right Accumulated management experience is a critical attribute of the transnational. This is why the Kenya government engaged Booker McConnell to set up the Mumias sugar production, processing and marketing complex. Access to relevant experience and skills can enable a developing country to avoid some common pitfalls by tailoring solutions to specific problems, making the most efficient use of available raw materials.

Market knowledge is a critical attribute in export marketing. The

opening for cassava imports into the E.E.C. was created by the familiarity with the animal feed industry there and active participation in it
of firms like Krohn, Toepfer and Peter Cremer.

Knowledge of the *quality and presentation standards* required to meet
competition in distant markets is another major transnational attribute.
A clear understanding of how bananas should look if they are to
compete in high-income consumer markets and of how they should be
handled to achieve a uniform unblemished appearance, has been a key
ingredient in the commercial success of Del Monte and United Brands,
the two transnationals that are the main banana exporters.

Ability to apply quality and presentation standards required elsewhere can also help in meeting domestic consumers' preferences as
they evolve. The standards brought by Unilever to margarine in
Turkey influenced both the domestic competition there and eventual
quality control legislation. The risk of contamination in the small-scale
manufacture and sale of ghee had previously been high.

Advantages to developing countries of transnational participation

Many of these advantages are evident from the aforegoing. With
foreign exchange acquired through their operations in other countries,
transnationals can bring in equipment and other resources for which
convertible funds might not be available domestically. Provision of
new capital leaves local capital free for other uses. Moreover, once a
transnational has brought in seed capital it may, as in the case of
Unilever-Is in Turkey, continue to generate new capital which is
channelled back into the business to fund new projects and future
expansion. This retained profit can easily outstrip the value of the
original seed capital and is therefore at least equally important in terms
of investment in a country's future.

Skills passed on A transnational may bring in some foreign managers
and technicians initially, but generally it will find it more economical to
train local personnel to carry on the enterprise locally over the longer
run.

Over the years, this continual investment in training builds up and
sustains a cadre of competent managers with the skills necessary to
operate modern businesses – skills such as long-term strategic planning, evaluation of market needs, assessment of risks, and an ability to
open up international trade and business opportunities. This kind of
training is generally much more effective than that available through
programmes divorced from continuing practical application.

A number of the employees trained by transnationals go on to join other local companies, or to work in national professional or training organizations. They may also enter Government service, and apply skills acquired during their business career to the management of their country.

Back-up in technology Firms like Unilever and Cadbury are able to provide a full range of support activities such as technical, engineering, safety, productivity, financial, data processing, environmental protection, personnel, training, legal and marketing services, which together enable their companies to meet the opportunities and problems of the particular environment in which they operate. In many countries continuing research on the production of the raw materials they need has brought important improvements in agriculture. Cadbury pioneered the development of cocoa production in India. It sponsored research, provided seed and advisory services to farmers, set up local buying stations and offered an attractive price to stimulate production. McIlhenny and Jamaica Broilers provide a continuing extension-type guidance to their contracted growers.

General economic benefits These commonly include contributions to foreign exchange earnings via increased sales on foreign markets or substitution of imports, expansion of agricultural production and income because of the additional outlets provided for farm products, provision of additional employment to developing country people and contributions to local and central government revenues via taxes, etc.

The foreign exchange earnings of Thailand from cassava exports grew in a few years from nothing to over $500 million, rivalling export earnings from rice. When Mumias came into full production it eliminated the need to use foreign exchange on sugar imports into Kenya. While Jamaica Broilers still imports much of its feed ingredients, medicaments, etc., the foreign exchange outlay is much less than if the country were to import frozen broilers. Additionally, 260 farmers are raising chickens profitably in Jamaica and many more people are employed in the feed mill and broiler processing and distribution.

The foreign exchange accruing to Honduras from exports of barrels of salted peppers at $753 000 in 1980 may not seem large but it was net. The farmers concerned were making a profit of $870 per acre in contrast to an annual national average income of about $500. Farmers growing sugar for Mumias in Kenya received payments well above the national average whereas previously, cash income had been rare; 14 000 new jobs were created. Under the influence of expanding demand for margarine, sunflower production in Turkey expanded from 50 000 to 700,000 tons with consequent benefits to the rural areas concerned. The extension of cassava growing in Thailand stimulated by expanding exports was of direct benefit to some of the poorer parts of the country.

Mumias was a major contributor to Kenya Government revenues, paying around $12 million annually in excise tax. Over the decade 1955–65 Unilever-Is paid $10 million in tax to the Government and $2 million to its national shareholder, taking out of the country only $8 million. Cadbury in India paid over the bulk of its net earnings to the Government in tax, repatriating to the U.K. only the dividend on its capital shares less income tax.

Support requirements

For the entry of a transnational enterprise into a developing country economy there must be a favourable attitude on the part of the Government and the people, and a fair expectation that this will continue.

The existence of favourable operating conditions, reasonable security, the availability of suitable labour and adequate infrastructure, e.g. reliable water and power supplies, communications and transport, is another requirement.

For low-cost operations, it may be crucial to be able to import on favourable terms equipment, supplies and raw materials that are not available in the developing country. Having to use more expensive and less efficient local substitutes can greatly increase costs, to the disadvantage of local producers, consumers and foreign exchange earning capacity on competitive markets.

It is also essential to be able to repatriate sufficient profits to provide an incentive for the capital investment, acceptance of risks and for the management initiative applied. Access to research findings, for processes under patent and for highly productive plant or animal breeding or processing materials such as special ingredients will also need to be charged for. This is implicit in the hybrid sunflower seed distribution enterprise initiated by Unilever in Turkey, for example. The external partner, the seed breeder, will require reimbursement in dollars for use of his know-how.

Where Governments impose price controls to restrain inflation, impact of inflation on costs must be recognized. Unilever did not suffer too badly from price controls during phases of accelerated inflation in

Turkey because the Government took account of costs over the whole cooking fat industry. Jamaica Broilers were in a similar position. In the case of Mumias Sugar Company the Kenya Government held down the wholesale price of sugar in the years up to 1981 to the point where the firm, although operating more efficiently, was almost bankrupted.

To bring in a transnational enterprise on favourable terms, some kind of intermediation may be required. Often there can be mutual interests which are not readily perceived by the participants. In the case of McIlhenny and the Honduran farmers it was a religious organization that helped make their contracts work. World Bank staff promoted negotiations with an American tobacco company to set up a factory in Paraguay. A commercial agency was engaged to seek external investors in a development area on the Mediterranean – with payment by results.

Issues for discussion

1. What is the status of transnational enterprise in your country? In which lines of commodity trading is it active?
2. Are there official restrictions on the operations of transnational enterprises in your country? If so what is their nature? What was the origin of their introduction? Are the factors associated with these restrictions still relevant?
3. Does the Government of your country have a policy of attracting investment by foreign enterprise? If so, what are the incentives offered, the security provided and assistance afforded with physical and social infrastructure, complementary local finance, etc?
4. Taking two transnational enterprises that have operated in your country for a substantial period of time, assess the benefits that have accrued to your country from their operations. Have there also been disadvantages? If so, attempt a net balance of gain and losses.
5. Assess the degree of competition faced by these two transnationals in the marketing channels in which they operate. What are the implications of your findings for the prices they pay to suppliers and charge to their customers? If there is an element of monopoly, what are its limits?
6. Assess the bargaining power of two different transnational enterprises operating in your country *vis-à-vis* the farmers who supply their raw materials and the Government. What are the alternative sources of income for the farmers and sources of supply for the

transnationals if they abandon their present operations? What would they lose in terms of investments committed, established markets, other assets including their public image?

7. Has the presence of a transnational enterprise in your country generated the growth of national competitors and extended use of the technologies it introduced. Have people who acquired training and experience working for it played a wider role in the national economy?

8. Identify for your country some areas of domestic or export marketing where the entry of a transnational enterprise could be beneficial. What are the gaps where new initiative, investment, technology, management, co-ordinated integration, external market linkages are needed? What should be the attributes of the transnational if it is to meet these needs? On what terms could the Government conveniently invite its participation.

9. Do you know of a situation where a transnational enterprise currently has a negative public image in a developing country, or has withdrawn or been taken over by the Government, after a previously successful operation for a substantial period? Identify and appraise the reasons for this. What could have been done to avoid this breakdown of relations?

5

Co-operatives

Ogbomosho Society, marketing food crops – Nigeria

Food crop marketing in Nigeria has generally proceeded on a small scale with many participants and transactions. Supplies originate from farmers scattered over a wide area each offering relatively small quantities. Contacts between sellers and buyers are predominantly personal: they are cultivated by selling on credit to trusted customers and by occasionally exchanging gifts. Once these ties are established it becomes difficult for one seller to entice away another seller's regular customer which tends to reduce competition. Nevertheless, there is still a large body of buyers and sellers who deal with each other on an impersonal basis.

In the market, prices are determined through a complicated haggling process which requires skill and experience. Each seller has his own assessment of a buyer's economic and social status (as revealed by dress, jewellery, etc.). There are also degrees by which a buyer is considered a 'regular customer'. Standard commodity prices may be established early on in a particular market, and then remain fairly uniform among all sellers. For commodities sold more rarely and those whose quality shows many ill-defined gradations, prices remain rather fluid.

The establishment of co-operatives to undertake food marketing in Nigeria has been suggested at various times. There is the general appeal of co-operatives based on their success in other parts of the world: they introduce new competition, they can improve the bargaining power of farmers and perhaps short-circuit some marketing links, so saving on marketing costs.

Despite these anticipated advantages over the private marketing system, most of the food crop marketing co-operatives set up in Nigeria have not had much success. The Ogbomosho Multipurpose co-operative society, however, has kept going for a considerable time. It was established along with many others when the Federal Government launched the Green Revolution programme. Its goal was to strengthen the agricultural sector and achieve self-sufficiency in food production. Credit was made available, in particular, to facilitate use of agricultural inputs by smallholders.

The Society is governed by its members who meet from time to time and is run by an elected management committee. There are sub-committees for market survey, propaganda and finance. The market committee assesses the quantities of food crops to be purchased by the Society, and sets their wholesale and retail prices. It can make reductions in the prices of perishable and slow-moving produce as necessary.

Marketing process

The first step is for produce to be delivered by a member to the Society. The secretary of the Society and another representative, usually the treasurer, weigh it and pay the member. The secretary, helped by a few co-operative workers, then puts it into storage or sells it directly to retailers.

Ogbomosho Food Crops Marketing Society accepts supplies from both members and non-members. It buys mostly from farms in its area taking up large quantities during the harvest seasons, and holding them in its store. Maize, cassava, yams, beans and rice are weighed, cleaned, but not graded specifically. Yams are stored on elevated planks for aeration. Maize, beans and rice are bagged. Mostly, the Society sells wholesale, but, to meet the demands of poor consumers, some crops are sold retail. There is no advertising or other promotion because the Society has no problem in selling its products. Decisions on quantities for sale, packaging, the marketing channels, the time and place to approach potential buyers – are all made by the market committee. Most crops are sold out after a few months of storage.

The Ogbomosho Society hires vehicles to carry the crops to the market: it has no vehicles of its own. During the peak season for maize and yams, this means hiring three to four motor vehicles for several weeks and hiring casual workers to load and off-load crops. The Society has customers who come from as far away as 50 km. It is noted for its high-quality yams.

The Society uses various weights and measures in accordance with local practice to price produce by the bag, the basket, by the heap for yams, and by bunches for plantains. Because of a lack of accepted grades, quality assessment is by subjective judgment, with examination by taste and touch.

Finance

To finance successive phases of marketing, the Society has two sources of funds. One of these is the Co-operative Bank which lends short- and medium-term to the Ogbomosho Society through the Union to which the Society is affiliated. The conditions of these loans include:

(i) Unions and societies can borrow up to ten times the amount of their share capital in the Bank.

(ii) A minimum share capital in the Society or Union must have been subscribed by the members.

(iii) The members of the Society must show that they are proficient in the work they wish to undertake.

(iv) In any project for which a loan is required, the Society or Union must provide at least 20 per cent of the capital required; the Bank will only lend up to 80 per cent.

(v) No loan is granted in excess of the Society's or Union's collective liability unless the excess is covered by the pledging of an amount on deposit in the Bank or the mortgaging of real property owned by the Society or its members.

The other source of finance for the Society is its member's share capital. This is cheaper than outside loans. To build up its own capital, the Society has embarked on a massive campaign to recruit new members .

Table 5.1. *Operational summary: Ogbomosho co-operative, 1980 and 1982.*

Crop	Quantity handled	Gross income	Handling charges	Net income
	tons	$	$	$
1980				
Maize	12	11 022	4 524	6 498
Guinea corn	16	11 022	5 035	5 987
Cassava	17	13 778	5 842	7 936
Tomatoes	4	7 274	1 433	5 841
Total	49	43 096	16 834	26 262
1982				
Maize	24	15 968	6 724	9 244
Guinea corn	26	17 193	7 123	10 070
Cassava	18	11 094	4 833	6 261
Tomatoes	6	6 294	1 701	4 593
Total	74	50 549	20 381	30 168

Operations

The main food crops which the Society handles are maize, guinea corn, cassava and tomatoes, see Table 5.1. From 1980 to 1982 tonnage sold rose from 49 to 74 tons and by value, from $42 000 to $50 000. In 1980, users of the Society were left with $26 000 as net income, and in the following years around $30 000.

The Society is also active in savings and loans. Farmers' savings amounted to $6460 by 1982. Loans to members amounted to $14 500 in 1982. The Society also distributes inputs such as fertilizers and insecticides. Farmers repay loans in kind or in cash.

Problems

Although the Society is operating relatively successfully and makes a profit, it faces a number of problems. Produce handling and packing follow traditional practice which results in considerable wastage of food crops. Most tomatoes are sold on the farm. Picking may be let out to a third party who has no direct interest in the price obtained. The Society is trying to help by having its agents supervise picking on the farm and grading at its store.

Membership is still low, since most farmers are not fully aware of the objectives and potential of co-operatives. The Society is therefore maintaining a propaganda campaign at the village level.

A serious problem is management: most of the paid secretaries, officials and accounting clerks are untrained and this has slowed down the progress of the Society. It would like to send its staff for training.

Membership illiteracy is also a problem. The Society has an educational programme on co-operative principles and practice and their limitations. The goal is to create a loyal, active and interested membership that will stand by the Society in a period of difficulty. A more enlightened membership would also be in a better position to check possible excesses by paid officials.

Benefits: direct and indirect

In spite of these problems the Society has benefited consumers, farmers and the economy as a whole. Its entry into food crop marketing has enabled consumers to buy at lower prices often without needing to travel to distant towns.

Farmers have their own marketing organization and feel that it belongs to them and that they can influence its activities. In addition to having an additional outlet for produce and a source of supplies competing with existing firms, its net surplus accrues to them jointly as capital or as patronage dividends. Its savings and credit facilities are also of direct benefit.

The Society has provided employment for a number of people and helped keep youths in the area who would otherwise have migrated to towns. When it started up, roads into the rural areas were repaired, various areas becoming more accessible so that farm produce could be moved out without risk of delay. The co-operative has been a direct link between the producers of the area and the Government.

Tsaotun Farmers' Association – Taiwan

The farmers' associations of Taiwan are multi-purpose co-operatives whose overall goal is to raise farmers' knowledge and skills, expand their output and income, improve their living conditions, and develop the rural economy.

The service area of each farmers' association is an administrative district. There is the overall provincial association, 22 county or city associations and 284 township or district associations. The township association is the backbone of the whole system. It is in direct contact with farmers; marketing their produce is one of its most important services. The Taiwan associations follow the traditional co-operative principles of open membership, democratic management and no discrimination over religion and political beliefs. They do not, however, redistribute surplus income to members as patronage dividends, but use it instead to provide extension and other services.

Organization

In 1981, the Tsaotun Association had 8598 regular members who were farmers and 2399 associate members. It had four operational sections: marketing, credit, agricultural extension and livestock insurance, and two administrative sections: general affairs and accounting (see Fig. 5.1). It also operated a wholesale market for fruit and vegetables.

The highest authority is the members' representatives meeting which elects board directors and supervisors. A general manager is engaged by the board of directors and is responsible to the board for the running of the Association's business activities. He appoints other staff and proposes operational policy to the board. Candidates for manager are generally recommended by the Provincial Department of Agriculture and Forestry for consideration and appointment by the board of economics, business or agriculture. The marketing section chief can be a graduate of a senior agricultural vocational school. Other marketing staff would have a high school or agricultural vocational education. The Provincial Farmers' Association (P.F.A.) can provide in-service training.

Marketing operations

The Association has two categories of operations: self-initiated business, and business entrusted to it by the government. Self-initiated business includes rice and fertilizers warehousing, marketing of various agricultural products, processing of rice and feedstuffs, supply of farm inputs and daily necessities, and operation of a fruit and vegetable wholesale market (see Table 5.2). For the first four of the above operations the marketing section is responsible. The wholesale market is under the direct supervision of the general manager with the approval of the county and agricultural authorities.

The Association has built 38 rice warehouses and four fertilizer godowns; total capacity is about 20 000 tons of paddy and 6 000 tons of fertilizer. Construction is mostly in reinforced concrete with a few buildings in traditional brick and wood. The Association met some of the cost, the bulk being covered by Government subsidy.

Fruits and vegetables are assembled, graded and transported to markets in consumption centres, particularly Taipei City. Pigs are

Fig. 5.1. Organizational structure of the Tsaotun Farmers' Association.

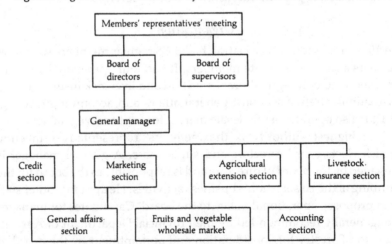

transported live to wholesale markets. These marketing services are provided for members on a cost plus commission basis.

The Association has been milling rice for many years. A large-scale rice mill was built with Government financial assistance. Present hulling capacity is 12 tons per hour, mostly working on paddy supplied by the Provincial Food Bureau (P.F.B.). A 16 tons per hour feed mill was built in 1964 to supply members with high-quality mixed feeds. The Provincial Farmers' Association provides materials for processing by the Association on a commission basis. The processing and sale of feed is the main profit earner of the marketing section.

Tools, pesticides, feed, seeds, are sold to members at cost plus a small margin, as are also clothes, soaps, sewing machines, radio and television sets, motorcycles, family electric appliances, food, etc. The Association is exempt from business and income taxes, and can offer prices a little lower than those of private shops. In 1982, a mini-supermarket carrying over 2000 items was opened. It has been warmly welcomed by both farmer and non-farmer members.

The fruit and vegetable wholesale market was organized by the Association and is the only one in the area. Business hours are 4.30 a.m. to 1.30 p.m.. Sales are by auction, which is considered fair to both parties. Purchase of the site and construction of the market were financed by the Association with subsidies from the government. Total sales in 1981 were 21 450 tons. The market is run by a separate board of directors with the general manager of the Association as chairman.

The Association also markets milk on a year-round basis and mushrooms from November to March. Milk is sent to the Provincial Farmers' Association dairy plant at Changhwa while the mushrooms go to

Table 5.2. *Main lines of business: Tsaotun Farmers' Association.*

	1979	1980	1981
	Tons		
Paddy stored	14 900	17 100	14 900
Paddy hulled	6 600	7 400	4 800
Fertilizer distributed	7 900	9 600	7 600
Feed processed	60 000	40 000	35 000
Fruit and vegetable market throughput	17 400	18 200	21 500
		$	
Sales of supplies and consumer goods	214 222	140 500	180 350

canneries under contracts. Prices are determined by negotiation between the two parties under the supervision of the government. For both these operations the Association receives a commission.

The collection, storage and processing of rice and the storage and distribution of chemical fertilizers are entrusted to the Association by the Government. The Association has been handling rice for the Provincial Food Bureau since 1953. Rice is received as payment of rural land tax, repayment of fertilizer and cash credits and to obtain the price guaranteed by the Government. Paddy is usually stored for a year or more then hulled into brown rice. The fertilizer is also stored and distributed for the Bureau. In Taiwan, the farmers' associations are the sole agents of the P.F.B. for fertilizers used on rice and upland crops. For this Government business the Association is paid a very low commission.

Marketing, credit and extension

Close co-ordination between marketing, credit and agricultural extension is an integral feature of the Tsaotun multi-purpose association. The marketing section needs finance from the credit section, which depends on the marketing section to collect repayments of production loans. The extension service helps both with production and marketing.

According to the Farmers' Association Law, finance for marketing operations may be derived from: (1) accumulated surplus, (2) advances from the credit section, and (3) loans from agricultural banks. The working capital of the marketing section of Tsaotun Farmers' Association comes from the first two sources. However, the limit on inter-sectional credits is $500 000. Fixed capital outlays are based on its own accumulated , inter-sectional credit, and subsidies from the Government. For an inter-sectional overdraft or credit the marketing section pays interest at the rate for unsecured loans set by the Government.

The credit section of Tsaotun Association is well managed and has operated with great success. Total deposits at the end of 1981 amounted to $24 million, the sixth largest among such associations in Taiwan, so it can give substantial financial support to its marketing section. According to Ministry of Finance regulations the maximum credit it may allow marketing and livestock insurance sections of the same association is 10 per cent of the total deposit or one million dollars. Currently, the Tsaotun marketing section borrows only $500 000, i.e. well below the ceiling.

For stock holding, the marketing section does not need much finance since the main products are financed by the Provincial Food Bureau. The marketing section would, however, like more working capital for its other sales and supply business.

Income and expenditure

Table 5.3 summarizes marketing income and expenditure for the years 1979–81. As it is a service organization, earning a high profit is not its main objective. The feed factory is the main contributor to the annual surplus. The P.F.B.-entrusted business always exceeded the other operations in terms of volume, but made little contribution to net surplus. However, the surplus from marketing achieved by Tsaotun ranked as the third largest in the country during these years. This may be seen as a measure of its success in taking up marketing operations additional to those for the Government.

Farmers in Tsaotun township have been able to obtain farm inputs, especially chemical fertilizers, of better quality at a lower price, delivered to their homes. They can sell their produce at the wholesale market which it initiated or through the co-operative marketing programmes. Either way they get a better price than they would otherwise. Co-operative marketing of pigs, fruit and vegetables has not been very successful, but it does provide farmers with a good alternative outlet.

The Association handles paddy and fertilizers for the Provincial Food Bureau. Without its help, P.F.B. would have to set up its own warehouses and rice mills and incur much higher costs.

Table 5.3. *Income and expenditure, marketing operations, Tsaotun Association, 1979–81.*

	1979	1980	1981
	($ 000)		
Marketing section			
Income	2933	3943	3405
Expenditure	2751	3848	3373
Net surplus	182	95	32
Fruit and vegetable wholesale market			
Income	110	132	167
Expenditure	94	115	125
Net surplus	16	16	42

Factors accounting for the success of the marketing enterprise are:
(1) The Association has had a sound organizational and financial base. Its managers have been competent and creative.
(2) Its new enterprises – the wholesale market and feed factory – were well adapted to an area of rice, vegetable and pig production, and gave the Association a lead in such business.
(3) Most farmer members are loyal to the Association. They do most of their business and deposit their money with it: they follow its advice in production and marketing.
(4) The former American–Chinese Joint Commission on Rural Reconstruction, the Department of Agriculture and the Co-operative Bank of Taiwan have given valuable technical and financial assistance.

At the same time the Tsaotun Farmers' Association has had its difficulties:
(1) The revised Farmers' Association Law does not make provision for the raising of capital so that the only funds available for investment and current marketing operations are accumulated surpluses. This handicaps the associations in undertaking new capital-intensive marketing activities.
(2) While the marketing staff of the Association are industrious, they lack professional training. The courses currently available do not meet their requirements.
(3) Some of the conditions that favour successful co-operative marketing are lacking in Tsaotun. These conditions include (a) specialized producing areas distinct from their major markets, (b) concentration and specialization of production, (c) homogeneity of production and output for market and (d) groups of farmers dependent on one or a few crops for their total income.
(4) The commission charges paid by the P.F.B. for the business done for it by the Association do not cover the actual cost. Indeed, the Association applies profits from other business to cover its deficit on government business. The unduly long periods it has had to hold P.F.B.'s paddy – often over 12 months – results in excessive storage losses and a low utilization of the facilities.

Ha-ee vegetable marketing group – Korea

Farmers' collaboration in joint cultivation or other endeavours is traditional in Korean rural society. It was an integral part of irrigated rice production. Mutual aid practices at the village level are still part of the agrarian culture. Now, the family unit is regarded as the most desirable form of farm organization catering to the inherent 'individualism' of most peasants. On a very small scale, however, it has shortcomings in both production and marketing, a higher cost and risk and a poor bargaining position on the market.

Commercially-oriented family units, facing the high opportunity cost of labour, have responded by organizing group activities to raise their incomes. Individual proprietors agree to grow the same cash crops at the same time and with the same cultivation methods. They buy supplies and equipment jointly and sell jointly-graded produce to pre-selected market outlets. These are called the Crop Unit Farming Groups or 'Jak Mok Bhan' in Korean. There are now more than 4000, of which some 300 are very active. The degree and range of their co-operation in production and marketing depends upon their leadership and infrastructural development.

Ha-ee village group

This village is located in one of the leading agricultural regions of Korea, in a paddy-field plain, about 12 km from Jeonju-Shi, a provincial capital, and 280 km from Seoul, the main urban market. 47 households farm about a hectare each. The major lines of production other than paddy are squash, lettuce, Chinese cabbage, cucumber and yellow melon. Squash grown under vinyl plastic in winter usually brings in one-third of the villagers' annual farm income.

Ha-ee was devastated during the civil war. In the late 1950s the villagers developed their farms on reclaimed land. About 80 per cent of the residents belong to the same family clan with the surname Hwang. The village is classed as a Welfare Village, the highest ranking in the Saemaul (New Village) movement. Communications are good. There is a bus to Jeonju City every five minutes; 27 villagers have their own telephones. More than half of the group members have completed high school. Two have been to college.

A horticultural association was formed in 1968 with 30 member-households to deal jointly with their production and marketing problems. The present Crop Unit consists of 43 member-households.

Average income increased more than four times during the eight years between 1974 and 1982 to around $900, almost double the national average. The Ha-ee Crop Unit has long had a campaign 'Let not land become idle for more than 10 days'. The double cropping ratio reached 384 per cent in 1982, more than three times the national average of 122 per cent. The Crop Unit members, individually and as a group, had 27 hectares of vinyl greenhouses and 3 hectares of tunnel-style vinyl-houses. Both facilities were used for vegetable production during the cold seasons.

The group began with the joint reclamation of 6.9 ha of river basin lands. Then it set up a small chopstick factory for slack-season labour utilization. In the 1960s the leader Hwang Se-Yeon introduced winter and cold season vegetable production under plastic. Joint purchase of supplies began in the 1970s, with co-operative grading and sale of produce. The group acquired its own pick-up trucks and began to use its bargaining power in contract sales and transport systems.

Organization of crop unit

Any villager can apply for membership: the main requirements are a co-operative attitude and type of farming. Membership is cancelled if a villager stops growing the produce adopted by the group or fails to abide by its regulations. Every member must pay to the Crop Unit two kinds of dues: 1) entrance fee of $130 per household and 2) operations fund contribution of 1–2 per cent of annual sales proceeds in each year. Any latecomer pays an additional amount equivalent to the amount in the fund divided by the number of members. In addition, the Crop Unit collects enough money to cover common costs such as store-operating expenses. The funds of the Crop Unit are used on educational programmes, group purchases and sales, mutual aid and loans and other common projects.

Officers serve for one year and can be re-elected at annual meetings. They are paid only their actual expenses on official duties. Their responsibilities are as follows:

Unit Head represents and operates the crop unit.

General affairs section responsible for general operations.

Guidance section technical and marketing extension services.

Purchase section buys agricultural supplies and daily necessities.

Sales section responsible for grading, transport and sales.

Information section responsible for technical production information and market/price movements.

Auditors supervise all business activities and performance.

There is an annual meeting, regular monthly meetings and irregular business meetings whenever needed. Unless otherwise resolved, the Head of the Crop Unit, in consultation with the appropriate officers, has the power to award contracts to merchants and transport agencies, allocate marketable amounts among members and hold back sales when needed. Proceeds of sales are received in the name of the leader. They are paid out to members on his instructions by the general affairs officer.

Production planning

At its annual meeting the Crop Unit decides upon the kinds, varieties, area and period of production of its seasonal crops. It takes into account i) other areas' production plans ascertained by field visits ii) seed sales of leading suppliers by region, iii) weather forecasts by the National Meteorological Observatory, iv) changes in market prices and consumers' preference, and v) changes in agricultural and trade policies.

During 1981/2 the Crop Unit grew 11 crops bringing in gross receipts of $830 000. Squash was the most profitable – helped by an informal agreement with a rival group at Tegu that Ha-ee would concentrate on squash and the Tegu farmers on lettuce. At one time the Ha-ee group had 80 per cent of the Seoul market for squash. Its share has since declined to 40 per cent under competition.

Farm supplies and equipment are purchased jointly from the district primary co-operative or another agreed source. Funds for this are mobilized from the Unit's own capital, loans from a primary co-operative, cash collaterals from merchants and transport agencies, and individuals.

Group marketing

Under the regulations of the Crop Unit, individual sales of designated crops are prohibited. (Previously, individual sales of produce at lower prices tied to merchants' loans had prevailed in this area.) All produce must now be graded and packaged to the Crop Unit's standards. A box of 20–22 kg of squash should contain (60 + 2) items, all washed, sorted, graded, labelled and packed. Any violation is first fined by an amount equivalent to one box, the second violation may result in loss of membership.

About 70 per cent of the produce goes to four wholesale markets in Seoul, where five consignee-wholesalers are exclusive handlers. The remaining 30 per cent is sold in the nearby Jeonju Horticultural Co-operative Marketing Centre and in other areas. The consignee-wholesalers are selected well before sowing and given a written contract on payment of a cash collateral. During the marketing season, their individual performance in terms of price received is compared weekly: the lowest performer will get only half as much produce the next week. If he agains shows the lowest price performance, he will cease to receive produce and his collateral money ($6500 each in 1982) will be forfeited under the contract. During the marketing season, a village representative is stationed in Seoul market in order to inform the Crop Unit quickly about market movements and prices. The Crop Unit has a policy of keeping its produce within a certain price range. Daily shipments are controlled accordingly.

One or two transport agencies are contracted to deliver Ha-ee produce. They must also deposit collateral money as a guarantee for their service – usually $7800 for a year. If they fail in timely and safe delivery, any loss resulting is charged against the deposit.

The Crop Unit has steadily upgraded the quality of produce offered and kept strictly to the most advantageous presentation, time of arrival and outlets for its produce. This has enabled it to bargain for a wholesaler's commission of 5 or 6 per cent as against the 8–10 per cent normal for smaller more variable lots. Similarly transport charges were kept down by making firms compete for a contract that amounted to as many as 4000 journeys per year.

Benefits of group farming/marketing

There are time and cost savings in joint growing of seedlings, purchase of supplies and marketing of produce. Superior bargaining power has brought higher prices. Table 5.4 compares individual and group marketing expenses for squash. Notice that the group saved 60 cents per

Table 5.4. *Comparison of group and individual prices and marketing costs for squash, 1982.*

	Group marketing	Individual marketing
	$ per box	
Price obtained	6.58	5.92
Costs		
Sorting and grading	0.46	0.53
Packaging and materials	0.54	0.55
Transport	0.26	0.66
Loading and unloading	0.05	0.05
Consignment fees	0.33	0.47
Total marketing costs	1.64	2.26
Net return	4.94	3.66

box of 60 by joint sales and obtained 65 cents more for it. This amounted to an average $800 additional income per member in 1982, when they marketed 1120 tons.

Factors in the success of Ha-ee crop unit

The following elements were important:
- strong dedicated leadership,
- common felt needs for a better livelihood through co-operation,
- mutual trust among members who are relatively young, highly educated and belong to the same family clan,
- steady improvement in farming technology by villagers as a group,
- rational decision-making procedures and farm management planning,
- honest and sincere trading relationships with merchants, and
- timely support from the district co-operative.

Without question, quality of leadership has had a dominant role. Hwang Se-Yeon came from the village originally and, because incomes were so low, he left for the city where he sold insurance on a house-to-house basis. He made some money, then worked for an advanced farmer growing vegetables under plastic some 140 km away before coming back to start up himself. He knew what the urban market required and insisted that all group-produce be graded and packed strictly according to instructions received from a wholesaler in Seoul. His main battle was with farmers, particularly older ones, who were reluctant to grade out produce not up to these standards. It was the younger, well-educated farmers who saw the benefits first and persuaded others to join. Hwang Se-Yeon has received an award from the President of Korea for his outstanding contribution.

Markfed fertilizer distribution – India

Of the 115 000 fertilizer retail sales points in India in 1982, 43 000 were co-operative. Nearly 40 per cent of all fertilizers was distributed through them. The Co-operative Marketing Federation (Markfed) of Punjab has played a historic role in helping that State achieve the highest fertilizer consumption level in India – 123 kg per hectare. It had 1973 member societies in 1982–83 with a share capital of $11.3 millions. However, the bulk of this was supplied by the Punjab State Government, only $600 000 representing holdings of member societies. Currently the Chairman of the Board of Directors, who is also the principal executive officer, is a State Government nominee.

Most of its functional managers have basic degrees in their respective fields plus a Master's degree in Business Administration. However, powers are delegated only to the extent necessary for executing the policy decisions of the Chairman or the Board. This may reflect a feeling that the activities of the Federation are constantly under the public eye; hence extreme care is needed to avoid a false step. For a number of years the Federation acted as an agent of the State Government in distributing essential goods including sugar, vanaspati and kerosene as well as fertilizer. It has also been involved in a range of agriculturally-related activities including buying and milling grain on behalf of the government.

Fertilizer marketing operations

Markfed distributes fertilizer from its own granulation plants, from other domestic producers and from imports. Its purchases are planned on the basis of:
- the quantity of fertilizer it sold during the preceding year; and
- the credits sanctioned by the co-operative banks to the village level societies who are the Federation's customers. Almost 98 per cent of sales by Markfed are against such credits.

Stocks are held in 120 depots with a total capacity of 150 000 tons. The average storage period is 45 days. The village level societies lift fertilizer from these stores in hired trucks. The Federation's role is to have available supplies for member societies as and when they require them.

The Federation and its member societies share a marketing margin established under the government controlled price and transport cost pooling system. On urea, Markfed gets $7.00 per ton and $11.50 on diammonium phosphate, the societies $8.00 and $9.50 respectively.

Credit arrangements

The co-operative credit sanctioned to farmers for purchase of agricultural inputs is disbursed by co-operative banks as:
- Component 'A': in the form of cash;
- Component 'B': as authorization to purchase inputs from co-operatives.

Before each crop season, the Punjab Registrar of Co-operative Societies determines the volume of credit to be made available under Component 'B'. Allowance is made for a 10 to 15 per cent growth in fertilizer use if funds permit. For the year 1982–83 the amount so fixed was $150 millions. This is then allocated to village level societies by granting them credit limits at the co-operative banks. On the basis of these limits the societies prepare bank drafts and present them to Markfed sales points.

Markfed itself generally receives two months' interest free credit on purchases from manufacturers or importer. Additionally it can obtain credit against stocks at an interest rate of 17.5 per cent from the Reserve Bank of India. A third source of working capital for stocking and distribution of fertilizers is a short-term loan from the State Government. Here the rate of interest is 6 per cent per annum.

Income and expenditure on fertilizer distribution account

Although fertilizer distribution was profitable enough in earlier years, between 1975–76 and 1981–82, the Federation has lost money on this account. Three factors have had a role in this.

Till 1978, Markfed was the only co-operative agency in Punjab engaged in the distribution of agricultural inputs. The Government then set up co-operative 'focal points' to provide additional and more convenient marketing facilities for farmers. The 'B' component credit was shared between two sets of co-operative organizations; Markfed received 60 per cent and the focal points 40 per cent. The role of Markfed in the distribution and sales of fertilizers was reduced correspondingly.

Secondly, the Federation did not have an effective system of inventory control. In the mid-1970s it purchased large quantities of potassic fertilizers which it was unable to move. There were other carry-over stocks that incurred heavy financing costs.

Thirdly, in the late 1970s the Punjab Government announced a subsidy of 50 per cent for superphosphate. The amount due to the Federation on this account was estimated at $6.3 millions, but only $3 millions has been paid.

Applying stricter inventory controls and other measures, the Federation broke even on fertilizers in 1982–83 and in 1983–84 showed a profit (see Table 5.5).

Specific benefits from Markfed's operations are:

(i) *Assured stocking and supply of fertilizers* The private distributor tries to avoid holding stocks during the off-seasons and is reluctant to serve areas inconvenient for transport. Through village level societies Markfed makes fertilizers available at the doorstep of the farmer.

(ii) *Holding the price line* In times of shortages private trade is alleged to raise prices. Markfed continues to supply fertilizers at official prices.

(iii) *Maintaining quality* As a farmers' organization, Markfed ensures
that the fertilizer it supplies meets with government specifications.
(iv) *Distribution of credit in kind* This is important in expanding
fertilizer use and farm output.

Overall appraisal

Markfed has had the advantages of state government support and
privileges. It also had competent and qualified professional managers.
Yet it was not able to keep up the tempo of its early years. One reason
for the recurring losses in the later 1970s may have been its rapid
expansion into fields that were not closely related, resulting in some
loss of management control and field supervision – specifically inventory control. Steps taken to rectify this are already paying dividends.

There are also clear constraints on managerial performance: the
responsible managing officer is appointed by the State Government
from its own administrative or co-operative service for probably three
or four years, then he will move on. This may help in retaining close
relations with the State government at various levels, but is unlikely to
give the Federation the business leadership and ability to respond
quickly to problems and opportunities. Knowledge that the top position will be filled in this way is also detrimental to the retention of high
quality career staff who aspire to eventual leadership of an enterprise
they have helped to create and know intimately.

Table 5.5. *Income and expenditure on fertilizer distribution: Markfed, 1980–81 to*
1983–84.

	1980–81	1981–82	1982–83	1983–84
	$ millions			
Expenditure				
Loan interest and bank charges	4.27	5.49	1.98	0.52
Transport	0.30	0.17	0.15	0.19
Warehouse rents	0.33	0.28	0.23	0.18
Cartage	0.18	0.07	—	—
Handling	0.24	0.21	0.16	0.16
Shortages, demurrage, rebagging	0.04	0.05	0.33	0.41
Promotion	0.22	0.04	0.03	0.05
Service charges	—	—	0.70	0.67
Total expenses	5.38	6.31	3.58	2.18
Margin recovered	3.29	3.68	3.60	2.95
Surplus/(deficit)	(2.09)	(2.63)	0.02	0.77

Coffee Co-operative Union, West Cameroun

Since 1958, arabica coffee producers in Cameroun's Western Province have marketed their output through six co-operatives under the leadership of the Union of Arabica Coffee Co-operatives of the West (U.C.C.A.O.). In recent years, about 100 000 smallholders have used the co-operatives to market about 18 000 tons of coffee and to purchase about 20 000 tons of fertilizer annually.

Both the co-operatives and U.C.C.A.O. fully control their own affairs, including finances and terms of employment of staff. Members of each co-operative elect a delegate assembly which, in turn, elects directors of the co-operative who appoint an executive committee and a chairman to manage its operations. These include the purchase of coffee at collection centres, where the coffee is weighed, graded, sorted, sacked and shipped to central warehouses for sale abroad. The co-operatives also distribute fertilizer and equipment, and administer seasonal credits which are used by members to finance the purchase of production inputs.

U.C.C.A.O. also represents the interests of the growers with the government, which sets coffee prices to producers and determines the amount to be paid from the price stabilization funds at the end of the crop year. U.C.C.A.O. also maintains vehicles and equipment and keeps the central accounts. Finally, U.C.C.A.O. manages the reserves accumulated from the difference between the f.o.b. price for coffee and the payments made to producers minus the operating costs of the Union. By law, about 20 per cent of these reserves must be retained as long-term protection for coffee producers' incomes. Most of the remainder can be invested in developmental projects with government approval.

The above summarizes the statement on the West Cameroun Co-operatives Union in the World Bank report, *Accelerated development in Sub-Saharian Africa: an agenda for action,* 1981. The Union became the counterpart agency for an I.D.A. rural development loan. Part went to strengthening its own processing, sorting, and other coffee handling infrastructure. It also became the implementing agency for training programmes and measures to improve food crop production in its area.

Financial position

Table 5.6 summarizes the operations of the member co-operatives of U.C.C.A.O. for the years 1978 to 1980. Each year they made a profit and paid out a bonus to their members. Over the three years from 1978 to 1980 the co-operative members of U.C.C.A.O. sold 39 000 tons of arabica coffee, 13 000 tons of robusta coffee, and cocoa to the value of $2 millions.

Firm prices and high margins for Arabica coffee contributed to a total net surplus of $20 million. These high surpluses, however, hid the losses on supply services amounting to $3.3 million, over three years which resulted from supplies being sold to members at wholesale prices, which did not cover the transport, finance and administration overheads incurred.

The running costs of the Union are financed by contributions from its member societies. These are:
- 1.5 per cent charge on export value of arabica coffee;
- 100 per cent of all electronic sorting expenditure;
- transport hire charges.

In 1979 the Union increased the arabica coffee handling charge to its members from 1 per cent of the world market price to 1.5 per cent. This increase, together with bank interest received on surplus funds, provided funds well in excess of what the Union required to run its services.

Whilst the period 1978–80 was generally successful for U.C.C.A.O. with significant surpluses accruing, there were grounds for concern over its future performance, viz;
- losses being made on robusta coffee handling;

Table 5.6. *Income and expenditure: U.C.C.A.O. co-operatives 1978 to 1980.*

	1978	1979	1980
	$ millions		
Income			
Coffee and cocoa sales	34.3	60.4	55.2
Other sales	3.0	3.9	4.6
Other income	2.3	1.7	4.8
Total income	39.6	66.0	64.6
Expenditure			
Direct costs, coffee and cocoa	12.0	26.0	29.0
Other direct costs	4.5	5.7	5.3
Staff overheads	1.2	1.8	2.4
Depreciation	0.8	1.0	1.7
Other overheads	16.4	22.1	20.4
Total costs	34.9	56.6	58.5
Surplus	4.7	9.4	6.1
Bonus paid	2.8	4.2	1.8

- neglect of cost controls;
- poor working capital management.

Robusta coffee is handled by the co-operatives acting as agents for a marketing board. For this they receive an agreed margin. Unsecured crop financing, poor coffee collection controls and old-style processing facilities can quickly eat up this margin.

While coffee prices were high in the late 1970s, U.C.C.A.O. did not pay very close attention to budgetary controls and cost accounting. Overheads constitute a high proportion of its costs which will lead to difficulties if coffee prices go down as they began to do in 1981. A high level of stocks meant high financing and other storage costs.

The working capital of the co-operatives was held in 13 separate accounts. While some were in credit, others were overdrawn and incurring unnecessary interest charges: More of these funds could have been deposited on a three-month basis so earning a higher rate of interest.

The consolidated balance sheet for U.C.C.A.O. and its member co-operatives together is summarized over four years in Table 5.7.

Table 5.7. *Consolidated balance sheet: U.C.C.A.O. and its member co-operatives, 1977–80.*

	1977	1978	1979	1980
	$ millions			
Assets				
Fixed assets	3.4	5.3	8.1	10.9
Current assets				
Inventories	2.4	7.6	10.6	23.4
Receivables, customers	8.6	11.2	32.1	12.7
Receivables, other	2.2	17.9	19.6	12.9
Bank accounts and cash	30.5	21.8	14.7	16.5
	43.7	58.5	77.0	65.5
Less short-term debt	7.0	18.3	29.7	14.7
Net current assets (working capital)	36.7	40.2	47.3	50.8
Total assets	40.1	45.5	55.4	61.7
Liabilities				
Capital	0.6	1.1	1.2	1.2
Legal reserves	20.0	39.3	42.7	51.3
Undistributed earnings	19.3	5.0	9.4	6.3
Long-term debt	–	–	2.0	2.7
Provisions	0.2	0.1	0.1	0.2
Total liabilities	40.1	45.5	55.4	61.7

These show a shift away from a very strong cash position in 1977. By 1980 inventories and receivables accounted for 80 per cent of total assets – stocks of arabica coffee in hand exceeded 10 000 tons. The increase in receivables reflected a slow rate of repayment by farmer members of credit received as fertilizers, etc. In 1979 the amount owed by customers was also very high – collections were slow and the average period of credit given amounted to six months with the interest cost borne by the Union. Slow processing of documents was responsible for this.

Quality control

   The achievement of the coffee co-operatives of West Cameroun was to assemble the output of large numbers of small growers, process it, eliminate the worst of the defective grains and waste material, and move it on to export channels. Because it dealt with small quantities brought in by many different growers, it had some difficulty in meeting competitive world market standards. However, it had a preferential position on the French market as an associate country of the E.E.C. Over the years, the coffee supplied by U.C.C.A.O. fetched prices 2 to 8 per cent lower than those of COOPAGRO, the export organization of 17 large-scale growers in Cameroun who had their own modern processing installations.

   In the early 1980s the Union bought an electronic sorter of which it was very proud. However, it replaced local labour and did not prove very easy to operate.

Personnel

   In 1981 U.C.C.A.O. and its member co-operatives employed 633 Africans and 10 Europeans: of the 77 African managers, 20 had completed a co-operative diploma course; 24 had some form of post-secondary school education. In relation to other co-operatives, U.C.C.A.O. had a better than average management team. Its accounting staff were not so strong: only four of its senior accountants had had any formal training. Salaries in U.C.C.A.O. follow a government scale for primary or agricultural institutions which is the lowest-paid category. Higher qualified accountants are able to obtain better salaries in the commercial sector.

   Much more staff training was needed. One of the features of the World Bank project was to build up a continuing training programme for U.C.C.A.O. beginning with field extension staff. Productivity was low among the smallholder coffee growers in the West Cameroun: their average yield per hectare of trees was only 300 kg, less than half of the 600–700 kg prevailing in countries like Kenya, Mozambique and Uganda.

Price structure

Producers receive a minimum price fixed by the Government, and a bonus depending on the actual market return and the percentage of good quality coffee grains. In 1982 the typical price paid to growers for a kilogram of green coffee delivered to a member co-operative of U.C.C.A.O. was 97 cents. Table 5.8 provides a broad indication of costs, margins and prices at subsequent stages of marketing through to the consumer.

Table 5.8. *Prices and margins: Cameroun arabica coffee exported to France, 1982.*

	Margin/cost	Price
	$ per kg	
Picking	0.14	
Pulping	0.05	
Washing	0.14	
Transport to co-operative	0.01	
Fertilizer	0.05	
Producers' net return	0.58	
Price paid to producer		0.97
Stabilization, development charges	0.32	
Processing by the co-operative	0.30	
Transport to Douala	0.30	
Port dues	0.04	
Storage and loading	0.06	
Commission	0.02	
Deduction for quality defects, delays	0.19	
Price f.o.b. Douala		1.93
Freight Douala – Le Havre	0.05	
Insurance	0.10	
Price c.i.f. Le Havre		2.08
Value added tax 5.50%	0.12	
Handling, storage, transport	0.07	
Roasting, grinding, packing	0.57	
Weight loss, etc.	0.47	
Distribution	1.52	
Price to consumer		4.82

The grower received about 20 per cent of the price at which his coffee, roasted and ground was sold to consumers in Europe. Of the grower price, 58 cents is estimated as net income and return to his investment. Part of the price that would otherwise be paid directly to the farmer is taken as a contribution to a stabilization fund, and to finance technical services to growers and other development projects.

This co-operative system is also seen as a convenient channel for subsidies to stimulate agricultural development. With a government subsidy, coffee seedlings are sold to growers for 1.4 cents each. A further subsidy is available for the uprooting of coffee trees after they have finished their economic life of 20 to 25 years, and for replanting. Sulphate of ammonia, imported at $10.00 per sack, is sold to growers at $4.40; however, because of the scale of subsidization the quantity available is limited and growers complain of difficulty in obtaining supplies.

This co-operative system has had a legal monopoly of the marketing of smallholder arabica coffee in the Western Province since 1962. In the absence of competitive pressure, the onus lies with the directors and with the management of the co-operatives to keep down costs, avoid overstaffing, and improve efficiency. One remaining criterion of performance in a monopoly marketing and service organization is the degree to which it retains its share of the market in which it sells, and how far the growers it serves maintain their output. In these regards, the U.C.C.A.O. co-operative system has remained quite steady.

Windward Islands Banana Growers (Winban)

Bananas are the main export crop of the four Windward islands in the Caribbean – St Lucia, Dominica, Grenada and St Vincent. Their market is the U.K. where they benefit from tariff and quota protection. However, they still face competition from other suppliers – Jamaica Producers and Fyffes, a subsidiary of the powerful transnational United Brands. For growers in the islands the U.K. is a distant market about which they know little. Acting alone they could never reach it. Winban has enabled them to secure a firm foothold in this market and derive a steady income from it.

Each of the four islands has its own Banana Growers' Association to which all export growers belong. Winban, the overall association, undertakes the following functions for the four island associations:

• organizes and co-ordinates the marketing of the islands' bananas in the U.K.;

- negotiates contracts and prices with Geest Industries Limited which provides marketing and shipping service;
- purchases inputs in bulk;
- undertakes research on agronomic requirements, pest and disease control, packaging, etc.;
- represents the interests of the islands at international meetings.

Figure 5.2 sets out this structure of marketing and associated responsibilities.

The island associations:

- select, purchase, box and transport bananas to the port of shipment;
- provide field support services to growers, including aerial spraying against leaf spot;
- receive and store fertilizers, boxing materials and other growers' input supplies and distribute them either for cash or on credit;
- receive payments from the shipper for members' bananas and make payments to growers;
- administer growers' reserve funds, e.g. a quality bonus fund.

Fig. 5.2. Windward Islands banana export marketing structure.

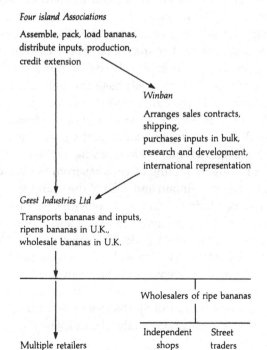

Four island Associations

Assemble, pack, load bananas, distribute inputs, production, credit extension

Winban

Arranges sales contracts, shipping, purchases inputs in bulk, research and development, international representation

Geest Industries Ltd

Transports bananas and inputs, ripens bananas in U.K., wholesale bananas in U.K.

Wholesalers of ripe bananas

Multiple retailers

Independent shops

Street traders

Ownership of the bananas passes from the grower to the island association at the boxing station. It passes from the association to Geest on delivery to Geest's depots at the port of shipment.

The price paid to each association is the same and is based on the 'green market price' agreed between Winban and Geest for the week of shipment. Deductions are made for handling and transport costs in the West Indies and in the U.K. including a standard allowance for shrinkage. The price to the grower is that received by the island association less a levy to cover: its services, the cost of materials and equipment supplied on credit, leaf spot control charges, hurricane risk insurance, export duty and a contribution towards the cost of Winban. This latter deduction is determined by dividing the Winban budget forecast by the production forecast for each island. Adjustments are made later according to the actual costs incurred and the tonnage shipped by each island.

Banana marketing is a highly capitalized operation involving an expensive infrastructure of shipping, ripening depots and distribution networks. Each stage from the time the bananas are harvested to the time they are distributed to retailers can be critical for sales quality. Careful organization and integration of activities is essential if bananas are to be competitive. Bananas are harvested to coincide with the arrival of the ships of the Geest line so that the minimum time is lost between harvesting and loading on board the ships. To retain freshness and minimum quality standards, bananas need to be cut a maximum of 48 hours before going into the cooled holds of the Geest ships. Shipments are generally made on a weekly basis throughout each year. During the voyage to the U.K. the bananas are stored under controlled temperature conditions of $56° - 57°$ F, at which ripening is held in check without damage to the fruit. After unloading at the port of arrival, the bananas have to be transported quickly to ripening depots.

The first contract for the marketing of bananas from the Windward Islands was made by Geest, Winban and each of the island societies in 1954. The critical role of Winban is to negotiate with Geest the 'green market price every week'. It is intended to reflect anticipated market conditions based on current market trends. This price is for fruit loaded that week in the Windwards, but which will not reach the market until two weeks later, allowing for the voyage and ripening period.

The 'green market price' is based on an estimate of the average primary wholesale price per box of all ripe bananas sold throughout the U.K. to secondary wholesalers and directly to multiples. This is based on an assessment of market intelligence gathered by weekly calls and visits to markets.

Organization of Winban

The Board of Directors consists of an elected representative from each island and a professional chief executive who is the Managing Director. There is an elected Chairman with a casting vote. Other professional executives are the Financial Manager, the Administrative Secretary, the Economist and the Director of Research and Development.

The head office of Winban is in Castries, St Lucia. Its research station is some 20 km away at Roseau. In view of the dispersal of Board members, implementation of board policy is very much in the hands of the professional executives under the Managing Director.

Improvements in banana handling

Until 1970, practically all bananas were shipped out of the islands on the bunch. Carrying or heading the fruit as bunches often results in physical bruising and damage. Winban's recommendation to growers has been to de-hand the fruit, preferably while it is still hanging on the plant. Handling and movement which can lead to mechanical damage is then minimized and the load carried is less, as the stalk is left behind in the field.

Winban designed a plastic field box to hold bananas cut from the stalk. The fruit was given better protection, quality was maintained and the reject percentage reduced. The boxes nest in each other when empty and cross stack when filled with bananas.

Winban Research and Development has also promoted field packing. This is being adopted increasingly by the more efficient and innovative farmers. The bunch is dehanded while still attached to the plant. As each hand is cut, its crown is immediately covered with a fungicide-impregnated pad. It is then packed in cartons on a table under the plant or in a small shed in the field. With this system, growers or groups of growers become more responsible for the quality of their fruit. Rejects are reduced and overall quality improved. For a grower to qualify as a field packer he must meet field husbandry standards specified by Winban and Geest: he must both deflower and sleeve his bunches. By 1983 40 per cent of the Windward bananas were field packed, mainly, however, on larger holdings representing less than 10 per cent of farmers. On average, the farmer-operated stations have produced better quality fruit than Association boxing stations.

There is more immediate motivation and responsibility for what happens to the fruit.

Evaluation

Around the mid-1970s a downward trend in production set in. Hurricanes, increasing input costs, and stringent export fruit quality standards discouraged some producers. Fluctuations in output upset the forecasting of market supplies. Competition from Latin American producers intensified. Concerted efforts were made to revitalize the industry through replanting and rehabilitation programmes. A pre-paid input levy was instituted: growers would contribute four cents per kilo from current sales of bananas to cover purchase of inputs for future use and so reduce reliance on the association for credit.

To improve fruit quality, field and boxing plant personnel training was provided, together with intensified quality assessments at the port. Research on quality was accelerated.

The organization of the banana growers of the four islands into associations with Winban as the apex body has been a clear success. Important factors in this success are:
  (i) The co-operative and democratic spirit of the island banana growers' associations;
 (ii) The research adapted to specific local needs undertaken by Winban;
(iii) The weekly marketing arrangements with Geest Industries Ltd.

Attributes, advantages and support needs

The basis of most farmer co-operation in marketing is to achieve economies of scale in transport and other services, and to raise their bargaining power over the price and other conditions of sale. By putting their small lots together, a group of farmers may be able to justify hiring a truck to transport them to a distant market where prices are higher. By all agreeing to use a store or processing plant they can justify the cost of establishing a better one than a single farmer or a small trader could.

By assembling their small lots and selling them as one unit, they save time and expense for a possible buyer and are likely to get a better price than if they sell separately. This kind of co-operation can be very simple – a group of farmers south-east of Hyderabad, India who save 20 cents per sack on their paddy by arranging for a transporter to pick up a full load at one place instead of each making his own arrangements separately, or a group of small farmers in Paraguay who send two representatives to negotiate with a buyer to take their produce from their farms at an agreed time and price. Without transport of their own, and situated along extended earth roads, waiting for a buyer to come to the farm would result in a much lower price than a negotiated group sale. The Korean farmers' group described in the Ha-ee case has achieved a remarkable reversal of bargaining power. Their sales weight was so strong that:

(a) transport agencies bidding for their business had to deposit caution money subject to forfeit if they failed to deliver on time; and

(b) commission agents seeking their custom had to cut their normal charge by several per cent, advance large sums of money as collateral, and accept a continuing competition between themselves to obtain the best prices under penalty of losing the business.

Winban and its four member banana growers' associations demonstrate the advantage of co-operative bargaining power in organizing sales on a very distant market. Negotiating as one organization, 20 000 small growers are able to arrange:

1) for a regular shipping service;

2) for a price based on independent observation of prevailing wholesale prices of bananas on that market.

Capacity to standardize quality of product, packaging and presentation over a larger volume of produce is another inherent attribute of co-operative marketing. This greatly simplifies the task of the subsequent wholesaler enabling him to offer a higher price or accept other favourable terms to obtain supplies from this particular source. The Ha-ee group profited greatly from strict grading and packing of their produce in accordance with central wholesaling requirements. Only by keeping up continuing pressure for this could the Winban producers match the appeal on their U.K. market of competing branded supplies. Such standardization must, of course, be in line with the requirements of the market served. For sales of food crops in the area served by the Ogbomosho society of Nigeria, no advantage was seen in departing from traditional practice.

Continuing concern for the interests of its members is assumed in the co-operative. Ability to count on a secure outlet means a lot to a farmer with many production risks to face. This holds good also for the distribution of farm supplies. While private fertilizer distributors in India might cut the price to secure advance sales, in times of shortage they could substitute inferior and unsuitable materials. Ability to count on a reliable service is one of the grounds for maintaining a co-operative through market phases when competitors seem more active and responsive.

Intrinsic attributes of the co-operative are democratic control and the return of profits to members as patronage dividends. These may be more significant in principle than in practice. Certainly, every co-operative must hold a meeting of its members annually and allow them an opportunity to express their views. However, a membership such as that of the Ha-ee group in Korea, over half of whom had been to high school and two had university degrees, would be very rare in most developing countries. Moreover, even in developed countries, farmer controls of a marketing operation that goes beyond the local assembly stage with which they are familiar can be a limiting factor. Ironically there are other aspects of democratic control that work in the same direction. Inflexibility and the lack of a capacity for quick decision-making (especially vital when markets change quickly or commodities are highly perishable) are often characteristics of co-operatives flowing from their democratic structure. Salary levels, are lower than those paid by private firms, which makes it difficult to attract or retain high-quality personnel. This is founded on attitudes toward management remuneration of farmer members conscious of their own low incomes. Similarly, with the management focus on service to farmer members and influential individuals among them, the funds available for patronage dividends tend not to be very large. Rather, the continuing handicap of many co-operatives, as voiced in the case of the Tsaotun farmers' association, Taiwan and Ogbomosho Society, Nigeria, is inability to mobilize capital needed for efficient operation and expansion. That co-operatives are not set up to make a profit has the legal advantage, however, in many countries of freedom from income and corporate taxes.

Advantages

Apart from the favourable tax position and some of the above attributes, the great advantage of co-operative marketing organizations lies in their popular appeal: to farmers, to governments and to aid agencies. Farmers like co-operatives because they believe they will obtain fair treatment from them. There is often deep resentment against traders. This was demonstrated in a recent study in Uganda of farmer attitudes. When asked to comment on the degree of honesty in business dealings, only 5 per cent of the respondents considered Asian traders honest all or most of the time. Fourteen per cent considered African traders thus. In contrast, almost 55 per cent felt their cooperative was honest all or most of the time. This does not mean, of course, that the farmer will necessarily sell through the co-operative if he can obtain better terms elsewhere.

Governments like co-operatives because they are convenient politically. Fitting in with ideas of traditional African forms of co-operation, marketing co-operatives have been seen by African governments as a way of realizing 'African socialism'. Just as African socialism was intended to be something different from the full-blooded Marxist version, so these co-operatives could be supported as a form of socialism not requiring direct state participation in marketing and having a strong democratic element. A co-operative structure of produce-buying and marketing can be fostered within a generally capitalist system just as in a socialist one, and does not require a political commitment from the government.

A second reason, more social perhaps than political, has been the desire to promote indigenous African enterprises and to replace dominant alien traders. This concern lies behind government promotion of co-operatives in many parts of Africa, also in South-east Asia where ethnic Chinese have often dominated marketing.

Co-operative structures may also constitute a convenient mechanism for combining the funding of rural development with political patronage. This is why in Malaysia at one time there were two rural co-operative systems, each a vehicle for promotion of a different political grouping. Co-operatives are a ready-made channel to rural areas for cheap credit and the politics of the party in office at the time.

There is also the view favoured by development planners that co-operative structures can integrate the various ingredients in rural development programmes – provision of production inputs and credit, demonstration of a new technological package and guarantee of a market outlet. The Ogbomosho co-operative was set up under such a programme. Similar thinking has also been behind the continuing government and international support to the coffee co-operative union of West Cameroun.

The appeal of the co-operatives to international aid agencies has two bases. For 'idealists' in donor countries an affinity towards 'idealistic' development approaches in the recipient countries is natural. They also tend to be biased against help to people and enterprises with

obvious profit motives. Secondly, there is the administrative convenience – for the unbiassed aid organizer – of channelling resources to
group organizations with appealing titles as against individual enterprises with uncertain qualifications.

Support requirements

Until co-operatives gain the experience and strength to compete on
equal terms with private traders, some special measures may be
required to assist them. These may take the form of concessional
finance for investments in marketing and processing facilities, preference in distribution of farm supplies, appointment as sole purchasing
agents of marketing boards, and exemption from certain taxes.

Ogbomosho society had access to credit on favourable terms. Markfed was the official distributor of fertilizers in the Punjab for many years
and continued to be protected by the allocation of specific quantities of
fertilizer for distribution by co-operatives and by its strategic link with
co-operative credit. The bulk of its operating capital was subscribed by
the State Government. Tsaotun farmers' association was well placed to
take up rice processing and the marketing of perishables, and to run a
wholesale assembly market, having been for many years the sole agent
of the Provincial Food Bureau in the purchase of paddy at a guaranteed
price and the distribution of fertilizer on credit.

Training facilities for co-operative staff are an essential support
requirement. The basic marketing tasks are complicated by the need
to keep accounts of individual member transactions, to comply with
government procedures and face up to pressures from individuals
pursuing their own interests. Because of the concern for co-operative
procedures, marketing management and ability to respond prag-
matically to changing market situations often come second in
many co-operative training curricula. The case studies from
Cameroun, Nigeria and Taiwan all stress the need for a better balance
of training and in-service courses adapted to specific marketing
requirements.

In most developing countries special government departments are
established to promote and support the development of co-operatives.
Commonly, there is a registrar of co-operatives who records officially
those societies that are entitled to tax and other privileges. Inspectors
are employed to visit co-operatives periodically, to check that their
accounts are in order and that members' interests receive due attention.
This is especially important where limited understanding of management procedure leaves the membership open to exploitation by

unscrupulous officers and local politicians. Continuing technical assistance will be needed in both management and marketing.

Issues for discussion

1. Is there an official policy in your country regarding the role of co-operatives in food and agricultural markets? If so, set out this policy and examine how it is applied in practice.
2. Identify the main co-operatives or co-operative federations in your country. How much business do they do? What proportion is it of the total volume of marketing in the sector concerned?
3. Prepare a profile of the marketing co-operative that is considered the most successful in your country. To what factors are its success attributed?
4. If there are some co-operatives in your country that are in difficulties, or are less successful than others, prepare an analysis of their problems. To what do you attribute their difficulties?
5. What is the capital status of the co-operatives in your country, or of the ones you know? Have they the capital they need to carry out the marketing functions they are undertaking? How do these co-operatives obtain access to capital and on what terms? Are changes needed in the legal provisions and other arrangements bearing on the capital structure of co-operatives?
6. Assess the management capacity of some co-operatives in your country. Do they fare better when an elected committee member takes the lead or where there is a paid manager with responsibility for marketing initiatives? How do the salaries paid to managers relate to those in other comparable enterprises? Is there any financial incentive for superior performance?
7. Delays in paying out to members the full price obtained for produce sold through a co-operative – less its costs – sometimes lead to members preferring to sell to other outlets. Does this occur in your country? If so, analyse the reasons for delay, assess whether it is avoidable and if so how?
8. It is often argued that there should be on-going co-operative marketing channels in parallel with private channels as a protection to farmers against collusion between the private firms and to maintain competition. Yet, where there is competition, private enterprises often tend to undercut the co-operatives. Identify and evaluate ways of keeping both channels in business without allowing either to have a monopoly.

6

Parastatals

BULOG: National food authority — Indonesia

The supply and price of rice has always been a critical issue for the Government of Indonesia as it is the preferred food of much of the population, has provided half of their calorie and protein intake, and been a source of income for over 12 million farm families. Production is uneven geographically and seasonally. Substantial redistribution of supplies over time and place has been needed to maintain a steady supply to consumers together with imports. The population of more than 160 million is spread over wide distances with difficulties in communication and transport. During the 1950s and 1960s regional shortages were frequent and retail prices soared. Large-scale trading in rice had, for a long time, been in the hands of ethnic Chinese. Policies to exclude them in the 1960s left a gap, with indigenous Indonesians still needing to learn marketing skills and build up capital and experience.

After the general political and economic disruption resulting from the attempted Communist coup of September 1965, the three major development objectives of the new government were stability, growth and equity. In the face of spiralling inflation it set out to eliminate the major causes. The fast rising price of rice, the basic food grain, was one of the most evident. It was in this frame that BULOG (Bureau of logistics) was established in 1967 as the single agency for rice purchases for the government and from 1968 as sole importer.

BULOG's objectives, as established over the succeeding period, were to:

i) stabilize supplies and prices to consumers in the larger urban centres to limit the disturbing effects of fluctuations on the cost of living;

ii) promote increased domestic production and improve the income of small-scale producers;

iii) meet the rice consumption needs of the armed forces and the civil servants;

iv) maintain reserve stocks to ensure food security.

It was also required to safeguard price stability for some other products such as maize, wheat and sugar. Since rice had by far the dominant role in BULOG's activities this study will concentrate on its rice marketing functions.

Rice operations

With the formulation in 1969 of a definitive floor and ceiling price policy for the most common rice varieties, the historic pattern of purchases to supply civil servants and the army at a minimum cost was reversed. Supplies would be acquired from the domestic crop when the farm floor price was threatened. Deficits would be covered by imports. BULOG support purchases ranged from 160 000 tons of milled rice equivalent in the drought year of 1972 to more than 2 million tons in the bumper crop year of 1981. Import needs fluctuated inversely with the domestic crop, requiring BULOG to organize its own crop surveys to forecast import needs earlier than was feasible via the traditional government crop estimates.

To bring the guaranteed floor price closer to farmers, BULOG's private buying agents were supplemented from 1973 by 2000 rural co-operatives (K.U.D.s). This required intensive marketing and management training of the co-operative cadres. With increased foreign exchange reserves after 1973, BULOG was also able to maintain large stocks and hold ceiling prices to consumers much more stable than during earlier years. Special task forces were established in the mid-1970s. They step in should co-operative collection and drying facilities become overloaded or private distributors hold rice off the market in order to profit from price increases.

Reserve stocks are seen in a stock flow perspective. Imports are counted as part of the reserve from the date of placing firm orders. Most imports arrive during the scarcity months before the major harvest. They are held off the market until needed to keep consumer prices within their ceiling prices.

BULOG releases food to the Department of Social Affairs for use in cases of emergencies such as earthquakes, droughts and volcanic eruptions. Supplies are also lent to village groups faced with crop failures to be repaid in kind after the next good crop has been harvested.

Table 6.1. *Rice operations of BULOG*

Year	Floor prices Cents per kg		Total production	Domestic procurement rice equivalent	Imports of rice	Stocks at year end rice equivalent
	Paddy	Milled rice	Thousand tons			
1970	5.7	10.0	12740	493	960	530
1971	5.3	9.4	13310	617	490	531
1972	5.0	8.9	12780	160	730	168
1973	7.2	8.9–10.8	14170	263	1660	579
1974	5.7–7.3	10.1–12.6	14810	530	1070	847
1975	10.0	16.5	14720	539	670	625
1976	14.1	23.4	15360	392	1280	541
1977	16.5	26.0	15180	424	1960	512
1978	16.1	24.9	16771	866	1840	1185
1979	12.0	19.2	17080	331	1930	783
1980	13.6–15.2	22.4–25.3	19270	1585	2030	1667
1981	16.6	27.7	21300	2014	525	2217
1982	18.0	29.3	22170	1836	300	1746
1983	14.8	23.5	23462	971	1160	1612

The operations of BULOG over the years 1970 to 1983 are summarized in Table 6.1.

The scale and direction of rice distribution by BULOG is shown in Table 6.2. Even with an excellent crop in 1980, it was still necessary to inject more than 1.85 million tons into the market. In part, this was because the low retail rice price desired by the government to temper other inflationary pressures created a large additional market demand for rice. Over the decade 1970 to 1980 rice consumption per caput rose from 110 to 134 kg. These quantities, added to the usual distributions, brought BULOG's total rice distributions in 1980 to a record of 2.75 million tons. Domestic purchases plus imports ranged around 10 per cent of annual domestic production rising to 17 per cent in a short year.

Organization

Figure 6.1 shows the broad framework of BULOG's organization, divided into three major functional groups: 1) procurement and distribution, 2) administration and finance, and 3) inspection and control. The 27 regional branches – called DOLOGs (Regional Logistic Agencies) – are organized similarly, communicating directly on technical matters with their functional bureau in BULOG and with the BULOG

Table 6.2. *Milled rice distribution by BULOG 1970–83.*

Year	Open market operations	Government servants including military	State enterprises	Social distribution and miscellaneous
	Thousand tons			
1970	229	711	151	37
1971	226	677	112	74
1972	419	651	84	57
1973	703	661	67	20
1974	315	658	112	133
1975	423	654	91	122
1976	888	662	89	77
1977	1703	635	80	100
1978	1224	586	92	107
1979	1802	660	101	49
1980	1859	639	90	74
1981	1183	684	92	44
1983	864	1396	93	43

Chairman for administration and policy. The 92 sub-DOLOGs, branches of DOLOGs, are located at district level with warehouse facilities and procurement and distribution staffs. In early 1983, there were 542 government-owned warehouse units plus 1394 units rented by BULOG from the private sector. The staff of experts and the Centre for Research and Development, with its technical experimentation and training station at Tanbun near Djakarta, all report directly to the BULOG Chairman. BULOG and DOLOG employment totalled more than 5300 people in mid-1982; of these 1700 were warehouse staff and 371 were college graduates.

Headquarters, regions and districts are connected by reliable telephone, telex and single-side-band radio, providing an all-important management information system. Daily reports of stock situation throughout the entire system are available at the touch of a computer key, with computer terminals located at the desks of key adminis-

Fig. 6.1. Organization chart of BULOG.

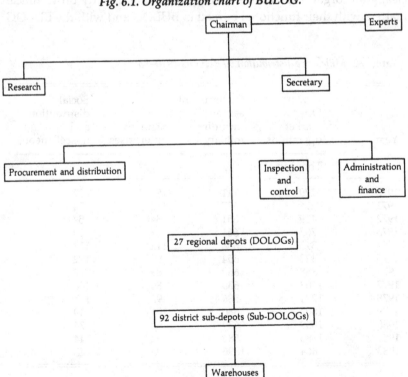

trators. Shortly, they will also provide daily finance and price summaries.

Intervention policy

This policy was developed to improve functioning of imperfect markets over widely separated areas and to stabilize consumer and producer prices at desired levels, with a minimum of direct marketing by the government. A farm floor price for common varieties is announced before the planting season begins. From 1976–77 price levels were designed to maintain a ratio of 2.2:1 between the increase in value of output and increase in cost of corresponding inputs. Ceiling prices reflect regional consumer purchasing power as well as distribution costs. The ceiling permitted in deficit areas recently has been 1.5 cents per kg higher than in self-sufficient areas and 3 cents per kg higher than in surplus areas.

The broad pattern of marketing channels to urban consumers is shown in Figure 6.2. Half the rice produced remains in the village. In the year 1981–82 39 per cent was bought by private traders: 11 per cent was delivered to the co-operatives at the floor price maintained by BULOG. It took up 8.7 per cent of total output, putting about half back on to the open market and delivering half to government departments (called Budget groups), etc. Their annual quantities are rather stable – around one million tons. These distributions have been maintained even though the inflation rate has been low in recent years, as they assure a partial annual turnover of stocks.

Large but variable quantities of rice are distributed through authorized private and co-operative (K.U.D.) distributors as necessary to keep market prices from exceeding desired ceilings. These market operations have ranged from 226 000 tons in 1971 to almost 2 million tons in 1980. Private traders tend to purchase higher quality rice at prices above the floor price or lower qualities at sacrifice prices, and to concentrate on transactions that can be completed quickly wihout much storage.

Marketing costs and margins for 1978, on rice bought by BULOG in the Krawang production area and sold in Djakarta about 70 km away, are shown in Table 6.3. By-products of milling remain with the miller. The subsidy amounted to 27 per cent of the retail price on paddy and 11 per cent on milled rice. Major cost items were interest charges and transport costs (averaged over all movements). In recent years BULOG's deficit on rice has been met from profits acquired as sole importer of wheat and soybeans.

Fig. 6.2. Rice marketing channels with percentage shares 1981–82.

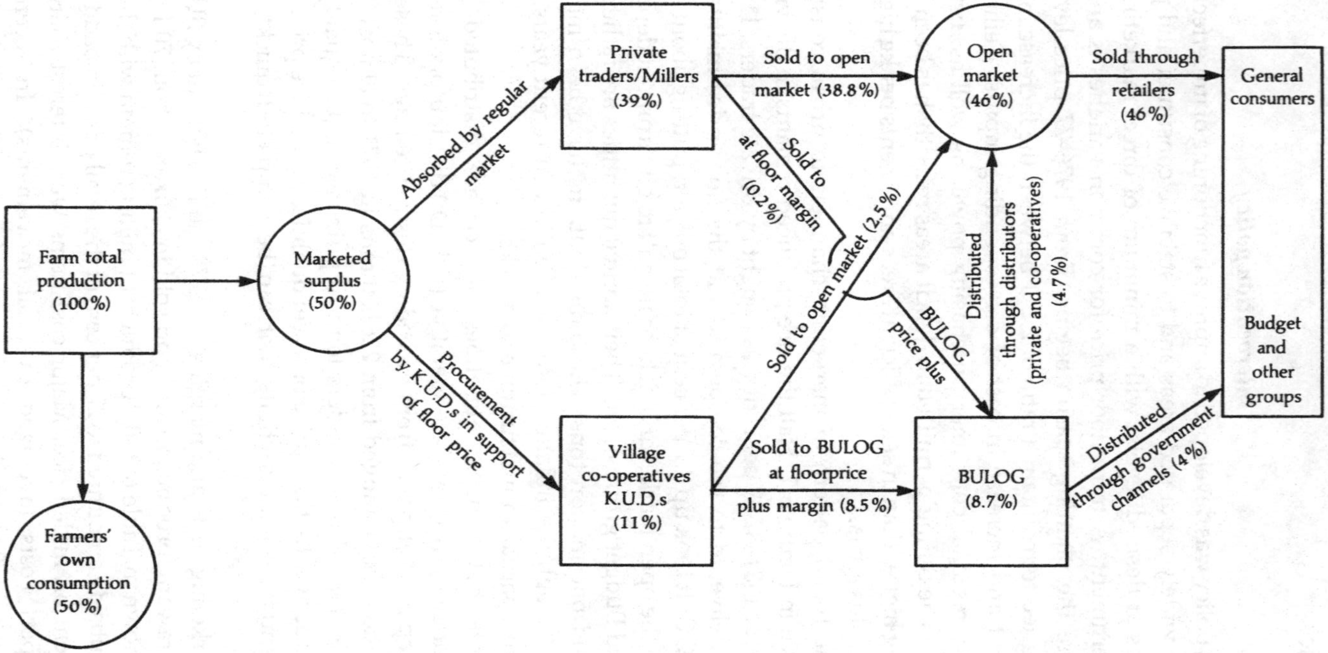

Table 6.3. *BULOG rice marketing costs and margins,*
April–October 1978.

(Rice bought at Krawang and distributed in Djakarta)

	Bought by DOLOG as paddy	Bought by DOLOG as milled rice
	Per cent of retail price	
Received at farm gate	83.9	83.9
Assembly to K.U.D.	1.2	1.2
Floor price to farmer at K.U.D.	85.1	85.1
Milling	4.2	2.9
K.U.D. margin	2.8	0.1
Cost to DOLOG at K.U.D.	92.1	88.1
Survey fee	0.4	0.2
Transport, handling	7.8	3.0
Movement between warehouses	2.8	2.4
Sack	6.0	2.9
Storage-paddy	2.3	–
Storage-milled rice	0.8	0.8
Losses	2.2	0.8
Interest and bank charges	6.1	6.1
Administration	2.7	2.7
Total BULOG costs/margins	123.4	107.1
Subsidy	(27.5)	(11.2)
Wholesale/retail margin	4.1	4.1
Retail price	100.0	100.0

BULOG financing

A sound financing basis is essential if BULOG is to support floor prices and hold sufficient stocks to keep retail prices from exceeding ceilings. It is able to take credit from the Central Bank up to the current value of stocks in the pipeline. To provide further security, the Finance Department of the Government pays directly to the Central Bank – from agency budgets – for the rice provided to government personnel and the army under contract.

BULOG does not handle cash in its supply and distribution activities. Payments for imports are effected by opening letters of credit for the supplier. BULOG's domestic procurement agents pay the farmers in cash. For this purpose, the K.U.D.s are financed by the Bank Rakyat Indonesia; private traders use their own funds. Upon receipt and survey of purchases, BULOG then pays its agents by bank transfers. Supplies are released to distributors only after evidence of prior payment at the bank to a DOLOG account.

The key to availability of bank funds at the right location at the right time for BULOG and its agents lies in careful logistical planning and transfers of BULOG and K.U.D. funds between banks by letters of credit in anticipation of needs. Similarly, for DOLOG routine or emergency operations, funds are transferred by letters of credit from BULOG's account according to specific budget allocations.

The success of BULOG

Management has been good. Capable executives were appointed from the army who listened to, understood, and generally found ways to implement the recommendations of their technical advisors. Even more important, these administrators worked closely with, and obtained the confidence of, government economic ministers.

By daily monitoring of markets in urban areas down to the sub-district level and by releasing supplies as needed, retail price ceilings were maintained. Success in maintaining floor prices induced farmers to use high-yield seed and fertilizer. The resulting production increase averaged more than 4 per cent a year.

A question may be raised, however, as to whether keeping down retail prices in the face of inflation has promoted the consumption of rice beyond the level for which it has a comparative production advantage in Indonesia.

The impact of BULOG on the private marketing system has been to maintain a consistent pressure for the contraction of margins. Over large parts of the country, farmers receive over 80 per cent of the price paid by consumers. A negative aspect of this is that the larger rice milling and storing enterprises are being driven out of business. It is no longer economic for them to hold rice in store for a long time. This is left to BULOG with its subsidy.

Kenya Tea Development Authority

The success of K.T.D.A. is widely recognized. In roughly two decades it has organized the planting of about 54 000 hectares of tea by some 138 000 smallholders. It has become a major processor and the largest exporter of black tea in the world. In the crop year 1980–81 smallholders produced 146 000 metric tons of green leaf. This was processed into

33 000 metric tons of made-tea, 85 per cent of it for export. This integrated production/processing/marketing enterprise has been the major factor consolidating tea as Kenya's second most important export commodity, and third largest source of foreign exchange earnings.

The concept began with the Colonial Government of Kenya in the 1950s. The goal was to bring African smallholders into production of a valuable crop hitherto grown only by foreign owned estates. The K.T.D.A. grew out of recommendations that it:

1) be a commercial enterprise under its own direction;
2) be financed from sources other than the government;
3) have full control over smallholder tea planting, processing and marketing.

This control was considered necessary if it were to secure high tea prices, low costs and become economically and financially viable. The technical package – growing, picking, processing and marketing tea – had already been tested in estate production in Kenya. The function of K.T.D.A. was to apply it to large numbers of small-scale growers.

In the early stages, the key element was planting material – K.T.D.A. owned the tree-stump nurseries. It could exercise 'quality control' of growers and the land planted to tea through its licensing power and supply of planting material. Leaf collection was also organized and supervised by K.T.D.A. through a network of buying centres. These centres were so located that no farmer would have to carry his leaf collection farther than 2.5 kms. After inspection, the leaf was purchased by the field staff and transported to the factory for processing. K.T.D.A.'s policy was to purchase 'two leaves and a bud', the best quality of leaf.

Marketing is by auction at Mombasa and London, and by negotiated sales to private buyers. Along with the estates, K.T.D.A. is required by the government to sell 15 per cent of its output at prices well below international levels to Kenya Tea Packers for domestic distribution.

In its report for the year 1980–81 K.T.D.A. announced overall sales of 28 million kilograms. The net average price received was the equivalent of $1.13 per kg. Growers received an average price of 18 cents per kg for 146 million kgs of green leaf. The margin taken for factory, marketing and production support costs amounted to $6 million. This is equivalent to 21 cents per kg of made-tea. Of the total return from sales of made-tea, the growers received about 80 per cent.

K.T.D.A. has been a prestigious place to work, and therefore has had a fairly good choice of staff candidates. It has favoured young, technically qualified and ambitious candidates. Only a few specialized technical jobs such as supervision of factory maintenance, design and repairs are held by expatriates.

K.T.D.A. was intended from the outset to be independent finan-

Table 6.4. *Balance sheet as of 30 June 1981, K.T.D.A.*

	$ thousand	
Fixed assets		13 162
Investments		
Tea factory companies	22 770	
Green leaf price reserve	2 197	
Factory finance	598	
Fertilizer finance	3 000	
Kenya Tea Packers	760	29 325
Current assets		
Stores	434	
Amounts due from factories under construction	21 254	
Debtors – K.T.D.A. managed factories	27 967	
Debtors – prepayments	8 285	
Cash	7 069	
Less export receipts due to factory companies	(3 087)	
	61 922	
Deduct		
Current liabilities		
Creditors	4 068	
Growers for green leaf deliveries	32 394	
Bank overdraft	4 751	
	41 213	20 709
Net current assets		63 196
Financed by		
Finance for agricultural operations		7 087
Finance for factory operation		45 083
Green leaf tea reserve	2 392	
Factory finance reserve	1 000	
Fertilizer finance reserve	3 337	6 729
Development account		3 069
Mortgages, etc.		1 228
		63 196

The items 'finance for agriculture' and 'factory operations' refer to loans from the C.D.C. $30 million, World Bank $12 million, O.P.E.C. and USAID $1 million each, Government of Kenya $8.5 million.

cially. The government undertook to develop and maintain roads in the tea-growing area. 15 000 farmers hold about 1.6 million shares in the 16 tea factories so far incorporated as public companies. This gives them an interest in K.T.D.A. performance and some voice in its operations.

K.T.D.A. management has consistently refused to insulate itself from the market, by building up large reserves, for example. It maintains a small producer price reserve – about 4 per cent of current (1981) payments to growers, see the balance sheet and income and expenditures accounts for that year (Tables 6.4 and 6.5). A Factory Finance Reserve is drawn on for factory construction in advance of external loan disbursements. A third small reserve finances fertilizer purchases against eventual payments by growers.

Despite rapid expansion, K.T.D.A. has kept down its costs benefiting from economies of scale once the basic structure was securely in place (see Table 6.6). Haulage unit costs, for example, were reduced as tea regions became more intensively planted.

The benefits have been clear – increased foreign exchange earnings and buoyant incomes in the tea growers' regions reflected in improved

Table 6.5. *Income and expenditure accounts, K.T.D.A., 1981 and 1978.*

	1981	1978
	\$ thousand	
Income		
Levy on tea sales	6 426	6 680
Factory management fees	1 989	1 823
Sales of planting materials	576	520
Factory design fees	826	35
Interest	549	85
Sundries including surplus	296	187
Total	10 662	9 330
Expenditure		
Planting material	429	372
Field development	2 011	1 242
Inspection and collection	3 911	2 231
Head office	4 144	1 772
Depreciation	1 938	672
Farm and training schools	(68)	(39)
Total	12 365	6 250
Surplus/(deficit) for the year	(1 703)	3 080
Exchange gain on foreign currency loans	564	—
Surplus/(deficit) transferred to development account	(1 139)	3 080

infrastructure and better housing. Average incomes from tea were around $250 in 1981. Since most growers would have other income as well, this indicates a level of living comfortably above that of the area.

Factors in the success of K.T.D.A.

These begin with its smallholder base. Consistently, it has produced higher quality tea than the commercial estates. High-quality plucking is difficult to maintain with hired labour. This advantage was backed up by effective quality control.

Staff and farmer incentives had, however, to be pushing in the same direction, and both management and line personnel had to be account-able for results. The Authority's structure and operating methods are designed to accomplish this. Grower returns are directly related to world prices realized for specific factories' output. Field staff perfor-mance is subject to grower pressure through district tea committees and K.T.D.A.'s board. Field staff and growers alike are answerable to the factories for delivering high-quality leaf – and factory managements are quickly held accountable for the quality (and hence the price) of the resulting made tea.

A first payment goes to growers – monthly since growing season for tea is continual – against their individual deliveries to leaf collection centres. In 1981 the first payment was 8 cents per kg.

The variable annual 'second payment', is geared to provide incentive for quality. It reflects not only deliveries by weight, but prices actually fetched for the tea from the factory at which the particular grower's leaf is processed. Each factory's tea is sold with identification on the papers and stencilled on the box. Since second payments are at the same rate for all growers delivering to a single factory, an individual might be tempted to deliver a greater weight of poorer quality leaf in the belief that he or she could benefit from the high-quality plucking of others.

Table 6.6. *K.T.D.A. unit operating costs.*

Year	Cost per grower	Cost per hectare
	Constant 1980 $	
1962–63	132	727
1970–71	116	331
1979–80	76	193

This is limited by both leaf-buying procedures and group pressures. At leaf collection centres, growers are often told to re-sort their bags of leaf before they are accepted for weighing. K.T.D.A. field staff constantly emphasize to growers the connection between plucking quality and second payments, and the damage to common interests if some individuals fall short. Weighing and recording procedures are kept as public as possible which also increases grower confidence in the probity of the collection staff.

In 1982 K.T.D.A. operated 27 tea factories as managing agent. Each factory unit is a separate financial undertaking with its own board of directors on which K.T.D.A. is represented. The factory makes a first payment to the Authority monthly on the basis of leaf delivered to it. Rates of payment are agreed with K.T.D.A. The surplus left after paying a dividend on the share capital (averaging 8 per cent) is passed on to the Authority for distribution to growers as a second payment.

Organization

The Authority is managed by a Board consisting of Ministry of Agriculture nominees, representatives of the Commonwealth Development Corporation, its main source of capital, and of growers elected regionally. The Chairman is appointed by the Minister. The General Manager (chief executive officer) is appointed by the Board. There are also growers' committees. These are used as a channel for explaining policy, discussing proposed changes, the allocation of planting materials, etc.

The Authority has three operational branches – Extension, Leaf Control and Factories. The extension staff are seconded from the Ministry of Agriculture, but are paid by K.T.D.A. and are under its authority. K.T.D.A. is considered 'élite' employment, carries various privileges, so high performance is likely to be forthcoming to prevent re-assignment to Ministry duties. An Assistant General Manager for leaf control supervises leaf officers each in charge of 36 leaf buying clerks, plus collection transport. A Chief Factory Superintendent supervises five Group Factory Managers each responsible for five or six factories with their own managers, assistants and trainees.

K.T.D.A. has been spared adverse government attention. The original C.D.C. funding agreement of 1960 had a condition that no export tax would be levied on tea over the subsequent 20-year period – the duration of the loan. Nor have there been any non-tax devices for reducing tea sales proceeds. The Government has resisted possible temptations to interfere in operational matters, though it did replace the general manager in 1980.

Within these strategic parameters the Authority has established an incentive and accountability system which is to a significant degree self-reinforcing. There is an incipient tension between different stakeholders – growers, field staff, factories, senior management – in which it is in each group's interests to evoke high performance from others, and in which each group also has some sanction or incentive power to do so. Output and efficiency do not, as is so often the case, therefore primarily depend on one particular set of actors, who may either be insufficiently motivated or see their efforts vitiated by other groups with conflicting interest, over whom they have little or no control. Instead, K.T.D.A. has developed the potential conflicts of interest to drive the institutional system rather than to obstruct it.

Zimbabwe Cotton Marketing Board

The Cotton Marketing Board (C.M.B.) was established under the Cotton Marketing and Control Act of 1969. Today it is one of four boards operating under the Agricultural Marketing Authority (A.M.A.). It is run by a General Manager who is advised by the Cotton Committee of the A.M.A. Membership is made up of A.M.A. and farmers' union representatives and the C.M.B. general manager. The Ministry of Agriculture sends an observer. The A.M.A. advises the Minister on the pricing and marketing of commodities handled by the four boards.

The Board's fixed capital, on which it pays interest, was provided by the Ministry of Finance. Fixed assets are written off gradually. Seasonal finance to pay the farmer for seed cotton and await eventual sale of the products is arranged by the A.M.A. It sells bonds and bills to the public and borrows from commercial banks. This procedure enables the four marketing boards to achieve substantial economies in financing costs.

The C.M.B.'s responsibilities include:

(1) the purchase and storage of all seed cotton grown in Zimbabwe;
(2) ginning the cotton and marketing the lint and cottonseed;
(3) ensuring an adequate supply of certified planting seed for all growers.

In carrying these out the C.M.B. is backed up by control legislation. All cotton growers must register with it and grow varieties determined by it. Large producers must adhere to delivery quotas to the ginneries. Grade standards have been established.

Zimbabwe cotton marketing system

This is characterized by a tight co-ordination between growers, the board, research and extension services and the Zimbabwe Cotton Corporation (Z.C.C.) which handles all exports. Cotton breeding is undertaken by the Cotton Research Institute on the basis of international market requirements. Annual meetings are held to review the fibre length and strength, maturity, fineness, spinning performance and uniformity of new varieties. When they are adopted, the Board selects growers to undertake multiplication. They grow for the Board which, in turn, distributes the seed the following season.

Training in production, pest control and harvesting is provided for growers, workers and extension staff. The agricultural extension service, the C.M.B. and the Zimbabwe Cotton Corporation also collaborate in explaining to farmers the grading system used.

Purchasing and processing: Planting takes place in October to November with the start of the seasonal rains. In January all large-scale growers are required to report to the Board on the area planted to cotton and a first production estimate is made. Large-scale growers make a second return in March indicating their likely sales. These data, together with estimates of smallholder production, are used to forecast the coming harvest. This enables the C.M.B. to set up its delivery quota system, ginning arrangements, and selling schedules well in advance.

Farmers deliver to the nearest ginnery. For growers in the more remote areas, there are transit depots where cotton is received and graded. Special cotton packs are hired from the board for a small fee. Each grower sews on to each pack a calico label marked with his registration number. This number is the sole identification of the cotton. On delivery at the ginnery depot, Z.C.C. graders allocate it one of four classes. The farmer is then paid out, through a computerized accounting system, normally within eight days of delivery. A sample of all bales set below the top priced grade is kept for a period to allow growers to appeal against the grading if they wish. The four grades are based on colour and cleanliness; they are designed to encourage appropriate production and harvesting practices. A cross check on the graders is made by experienced lint classifiers who visit the depots on a frequent but random basis during the buying period.

Once the cotton has been graded for payment to the farmer, a strict quality control system comes into operation. Each bale is classified into one of about 40 'stack' numbers by appraising its fibre length, strength, fineness and colour. It is then stored in stacks consisting only of bales with identical stack numbers. This system is unique to Zimbabwe. It is the ability of the C.M.B. through this system to produce lint of consistent and specified quality that enables Zimbabwe to achieve excellent prices for its export cotton.

Export: Zimbabwe Cotton Corporation is successor to a private firm that was the original exporter of cotton from the country. In 1966 this firm merged with Anderson Clayton and Company, world leaders in cotton marketing services. Following a vigorous programme of training, Z.C.C. is almost entirely locally staffed. It is the Board's export agent. Fully equipped with laboratory and technical facilities its offices are linked on a 24-hour basis to an international network of brokers and importers.

Z.C.C. staffs and operates the cotton grading depots throughout the country. When a spinner sets out the characteristics of the lint required, the Z.C.C. can identify a stack of seed cotton likely to provide it. This is then ginned; the lint is baled and sampled. The samples are then checked at the ginnery and at the central quality control laboratories in Harare to determine whether they meet the requirements of the contract. This system meets much tighter quality specifications than those employed in many other countries where the lint is classified only after ginning and ends up more variable in quality.

The typical customer of C.M.B. is producing a high-quality, fine strong yarn which is usually 100 per cent cotton. This yarn is then woven or knitted into high-quality fabrics using fully automated capital intensive equipment. Such customers require, and will pay for, lint of consistent and known characteristics. The C.M.B. has located a specific market segment for its product and does not compete against the much larger output of such countries as the United States and the U.S.S.R. The entire marketing system – grower, researcher, extension worker, buyer and exporter – is designed so as to meet the requirements of this specialized market.

Cotton seed is sold to the domestic oil-expressing industry. It contains about 20 per cent edible oil and provides over half of Zimbabwe's supply of edible oils. The residue, cotton-seed meal, is a valuable protein for livestock.

In 1981, a very good year for cotton, farmers sold 195 500 tons of cotton to the Board and received $110 million. Roughly 60 per cent of the cotton came from 750 large-scale growers and 40 per cent from some 42 000 smallholders. Their cotton achieved a marginally higher average value of 54.7 cents per kilogram due to the more careful handpicking and sorting done by family members. Returns from the various markets and the costs incurred by the Board are set out in Table 6.7. In addition to the cash surplus, the Board retained 5000 tons of cotton seed for the next crop.

Growers of cotton for seed are geographically dispersed as a precaution against losing the seed crop through drought or pests. Field staff check and rogue all seed crops during the season and ensure that growers meet the standards of management laid down under the Seeds Act. The seed cotton is delivered to the Glendale depot of the Board and stacked in special areas. Before ginning, and delinting, all machinery is thoroughly cleaned. Samples of the delinted seed are tested for germination in the government Seed Testing Laboratory. The average germination achieved is around 90 per cent although the minimum standard is 70 per cent. Seed not meeting the required standards is sold for oil.

C.M.B. and the Zimbabwean economy

Cotton is of special importance in Zimbabwe, notably for its smallholders. It is the best crop for the more remote and marginal rainfall areas. A typical smallholder planting two or more hectares annually (he could manage five) can earn a profit of nearly $300 per hectare from cotton. The expansion of the cotton industry and the development of the neglected smallholder areas of Zimbabwe are integrally linked.

Table 6.7. *Summary operations; Zimbabwe cotton marketing board, 1981.*

	Total	Per ton
	$ millions	$
Payment to farmers for 199 500 tons raw cotton	79.7	39.9
Sales – lint for export, 56 000 tons	84.0	150.1
– lint for domestic mills 15 000 tons	16.8	112.3
– seed to expressors, 120 000 tons	12.6	105.0
Total sales	113.4	
Gross margin	33.7	
Costs – processing, baling, storage, administration, field and financing	14.2	
– freight and marketing	12.6	
Total costs	26.8	
Surplus	6.9	

The C.M.B. is not a parastatal with a large staff undertaking a wide range of functions: its approach has been to control the key points in the marketing system and to co-ordinate the activities of the other agencies involved. Thus, through its monopoly of certified cotton seed purchase and distribution, C.M.B. controls the varieties of cotton grown in Zimbabwe; it does not, however, grow the cotton seed itself. As the sole buyer of cotton, it ensures that all cotton lint offered for sale from Zimbabwe has been purchased and processed according to known and laid-down standards but marketing is through another specialized enterprise. Agronomic research is undertaken by the Ministry of Agriculture under guidance from the Board, the growers and the Zimbabwe Cotton Corporation. The Board does not engage in production extension and credit programmes, but it is active in marketing extension particularly in regard to its grading and quality control systems.

Selling cotton in the market segment within which Zimbabwe operates, is highly competitive. Use of a private firm as broker enables the Board to tap the flexibility of private enterprise within the policy environment of a state corporation. It has been argued that the use of more than one broker could be advantageous. The view of the Board is that, if more than one broker were involved, it would have to take over the operation of the grading and quality control system. Yet the brokers would still want to do their own testing for quality contracts. Such a duplication would certainly increase costs. Under current arrangements, the Z.C.C. has a clear incentive to maintain the integrity and efficiency of the grading and quality control system. The more reliable the system, the more straightforward is the sale of the lint. If grading and quality control became the responsibility of a third party, such as the C.M.B., the direct link between quality control and selling would

Table 6.8. *Cotton production and exports, Zimbabwe, 1965, 1975, 1980.*

	Value of production	Exports	
	$ million	Tons	$ million
1965	1	—	—
1975	46.4	33 000	27.6
1980	58.8	54 000	59.8

be lost. Zimbabwe could then lose its competitive edge in the high-quality high-value yarn market and have to sell in the overcrowded market for average- to low-quality grades. The farmer, and the country, would be poorer in consequence. The cotton industry employs 45 000 Zimbabweans. Foreign currency earnings from cotton lint account for about 7 per cent of the total (see Table 6.8). The textile industry also exports – earning $18 million in 1980.

In summary, the Zimbabwe Cotton Marketing Board has developed a remarkably successful marketing system. It is sensitive to the needs of growers and of the market and is designed to ensure that farmers reap the maximum financial returns from cotton sold to the Board.

The Zimbabwe Cotton Corporation has since been taken over by the Cotton Marketing Board. Direct sales agencies have been established in Switzerland and the U.K.

Botswana Meat Commission

The 'pride of the African beef industry' is how the Botswana Meat Commission (B.M.C.) was described in an article in *Agribusiness World*. Few who know the industry will dispute the judgment. Over the past 30 years the B.M.C. has grown into a sophisticated international business with a turnover of more than $100 million a year. It has not, however, been without crises. Supply has been interrupted by periodic drought and foot-and-mouth disease; markets have been disturbed by world over-supply and political events.

The Commission has its origins in a private slaughter-house, known as 'Lobatse Abattoir'. In 1958 this was acquired by the Commonwealth Development Corporation (C.D.C.); in 1961 additional capital was provided through a Livestock Producers' Trust and by the government. The Meat Commission was established by statute in December 1956. Agreement was reached with C.D.C. to purchase its 50 per cent holding and convert it to a loan. Thus the abattoir became wholly government-owned and regulated by the B.M.C. Law.

The purpose of the Commission as set out in the B.M.C. Law is: '. . . to secure that so far as is reasonably possible all livestock offered or available for sale in Botswana to the Commission are purchased and that the prices paid therefore are reasonable'. No slaughterhouse may be licensed for export unless it is operated by or in association with the Commission. This monopoly also includes the export of live cattle.

The Commission consists of a chairman and ten members appointed by the government. Four are selected from a list submitted by the Livestock Industry Committee, which represents the producers. Throughout most of its life, decision-making in B.M.C. was centralized under a general manager. In 1977 an Executive Chairman, Mr Titus Madisa, who had been both Chairman of B.M.C. and Permanent Secretary, Ministry of Agriculture was appointed. Under him, there were general managers for administration, finance and technical matters.

The assets acquired from C.D.C. were paid for over a ten-year period at 8 per cent interest. Most of the present investment has been financed by B.M.C. from undistributed revenues. A development allocation of around a dollar per head of cattle purchased accumulated $2.5 million by the end of 1982. This has been supplemented by fairly generous provision for asset renewal.

Over the 15 years to 1982 throughput more than doubled from 100 000 to 230 000 head of cattle. Annual payments to producers rose from $11.3 to $53 million. The B.M.C. has been able to pay high prices to producers because it has aimed at high value markets, has tried to maximize utilization of the animal, and, above all, has operated efficiently. It has also been a major contributor of tax revenues to the government.

Livestock supply channels

B.M.C. would not have been able to develop as it has without an efficient internal livestock marketing system. Botswana is a country of some 600 000 square km. Much of it is semi-arid with an annual rainfall ranging from 150 mm in the south-west to 700 mm in the north-east. The human population has grown from 500 000 in 1960 to 850 000 in 1983. During that time, according to official estimates, the cattle population increased from 1.3 to 3.4 million. The principal means of communication is the railway which runs through the more heavily populated eastern part of the country, connecting Lobatse to Zimbabwe and South Africa. Only from the early 1970s was there any hard surface road outside the towns. Lobatse is in the extreme south-eastern corner of the country, implying very long journeys for many of the animals slaughtered there.

Tribally-owned land accounts for about half of the total area. Some 4 per cent is freehold. The rest is state land, dry and with hardly any livestock. By custom, all members of the tribe have access to the tribal grazing land. Ownership of livestock, however, is highly skewed and many own none at all, particularly cattle. The larger herds of cattle are usually kept at a 'cattle post' many kilometres from the owner's home. Traditionally, fencing is not allowed in tribal areas but this is changing under recent government policy. The freehold farmers have access to capital and modern management techniques and exert an influence on production and policy out of proportion to their numbers.

Livestock offtake is high compared to many African countries. This is attributable to the high prices which B.M.C. has paid, a relatively efficient system for getting livestock to the abattoir, an increasingly commercial attitude on the part of the tribal producers, and the freehold sector. Slaughter for local consumption is estimated at 15 to 20 per cent of B.M.C. offtake.

The structure of the cattle marketing system in the 1970s is shown in Figure 6.3. Many of the freehold farmers grow on or fatten animals bought from tribal farmers as well as raising their own. Government finishing ranches have been established to provide a parallel service, but this is still small.

Speculators buy cattle from tribal producers for resale to freehold farmers, or to the B.M.C. or for grazing on their own land, depending on the market. In all the tribal villages there are traders running a general store who are required by their licence to buy cattle at their store, but cannot buy elsewhere.

In order to maintain a steady day-to-day throughput at the abattoir B.M.C. operates a quota system. Each supplier who wishes to send cattle for slaughter has to apply for a quota and is penalized for

Fig. 6.3. The cattle marketing system in Botswana.

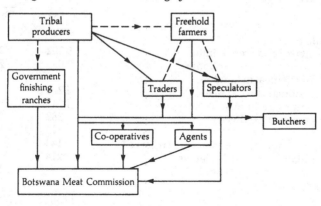

Immature _ _ _ _ _ _ _

non-fulfilment. Agents, have established themselves to help arrange quotas, organize rail transport and look after the producer/clients' interests at the abattoir. They operate on commission at competitive rates. Agents' touts persuade small producers, wishing to sell only one or two head, to get together to make up a shipment. A co-operative union acts as agent for the co-operative societies who have found livestock marketing an excellent basis on which to establish themselves.

B.M.C. purchases livestock by deadweight and grade. Grading is carried out independently of the Commission by graders employed by the Veterinary Department. In order to enable producers to plan ahead, the B.M.C. publishes a schedule annually of prices for each grade for each four-weekly period. At the end of the year any surplus is distributed as a bonus to suppliers (see Table 6.9). A deficit is met from a stabilization fund. Because the bonus goes to the person who sells the animals to B.M.C. (who may not be the producer), the

Table 6.9. *Income and expenditure accounts: Botswana Meat Commission, 1969 and 1982.*

	1969	1982
	\$thousand	
Income		
Sales of meat, meat products, services	14 343	102 190
Interest receivable	35	52
Increase in stocks	176	5 444
Miscellaneous	10	665
Brought forward	17	198
Total	14 581	108 549
Expenditure		
Payments to producers – initial	9 302	52 971
– bonus	1 095	7 833
Purchases of meat in London	—	767
Freight, storage, levies, sales	1 276	17 366
Processing and administration	1 343	15 637
Depreciation of fixed assets	382	1 356
Taxes	826	9 275
Transfer to capital loan redemption reserve	143	249
Appropriation to capital reserves	214	396
Transfer to stabilization fund	—	1 849
Carried forward	—	850
Total	14 581	108 549

Commission tries to keep down this part of the payment. Prices are occasionally adjusted upwards during the course of the season. The highest prices are normally offered from October to the end of February, in order to encourage producers to keep the abattoir supplied during the slack period.

For those suppliers living close to the line of rail this is the preferred mode of transport. Some deliveries are made by truck, but these are limited by a poor road system and disease-control requirements. Many animals walk part of the way and some arrive at the abattoir on foot. There are a number of trek-routes with watering points at convenient points.

Product marketing

To compete successfully on the world meat market calls for a very considerable expertise. Hygiene and disease-control regulations restrict access, limiting the extent to which products can be shifted from one market to another. It is a highly competitive market with larger-scale suppliers such as Australia, Argentina and, recently, the E.E.C. It is subject to over-supply from time to time as producers re-stock and de-stock in response to price fluctuations. Some markets are heavily protected by non-tariff as well as tariff barriers. The product is highly perishable.

The policy of the Commission has been to do as much processing as possible within Botswana while at the same time offering a high value product. Thus, at one time thousands of cattle were exported on the hoof. This was stopped after 1967 in order to retain the 'fifth quarter' within the country. A related policy has been to utilize and market as much of the animal as possible: edible and inedible offal, bloodmeal, bonemeal, carcasemeal, tallow, horns, hooves, ear and tail hair, ox gall all find a market. Hides once exported in a raw, salted form are now exported as wet-blues.

The basic market was for carcase beef in South Africa and other nearby countries accessible by rail. Its limitation is that climatic conditions are similar, so, when Botswana has a lot of beef to sell, the South African market is also glutted. Corned beef and frozen boneless beef were sent to overseas markets.

In 1973 a decision was taken to balance all operations to cope with a kill of 1200 head a day. This programme included a new freezer tunnel, new carcase chiller block and a new by-products plant. In 1976 the de-boning room was expanded so that, if necessary, the entire kill could be de-boned. The capacity of the slaughter-hall and by-products plant was later increased to 1800 and a new cannery was built. Pet food canning was added in 1981.

Meanwhile Britain had become a member of the European Common Market. Hitherto, Botswana beef had enjoyed Commonwealth preference. Now it would face a 20 per cent tariff and variable levies. During 1974, while negotiations were proceeding, disaster struck. The world beef shortage of two years earlier had become a glut. The E.E.C. closed its borders, first of all to frozen, then to chilled beef. Botswana's most important market, which in 1973 had absorbed two-thirds of its output, disappeared virtually overnight. Intensive representation eventually resulted in an agreement that Botswana along with Kenya, Madagascar and Swaziland, would pay only 10 per cent of the E.E.C. levy. Botswana was granted a quota of 21 000 tons of boneless beef on the E.E.C. market. The B.M.C. abattoir meets E.E.C. standards and important outlets have been established in Germany, Holland and France and by-products find a wide market. Efforts to develop further markets in Africa have been hampered by transport difficulties, and by foreign exchange shortages in recipient countries.

After many years of discussion with the government, it was finally decided that B.M.C. would establish a tannery in Lobatse with a British technical partner. It came on-stream in 1978 with a capacity to tan 600 hides a day to wet-blue stage. This was later increased to match B.M.C.'s full kill of 1200.

A second abattoir has been built at Maun in the north-west of Botswana with a capacity of 80–100 head per day. It can produce chilled carcases or frozen boneless beef.

Sales in Europe and South Africa have been effected by local agents. 'Export and Canning Company (Pty) Limited' (E.C.C.O.) was set up in 1961 in partnership with the Hurvitz Company to sell beef and offal. This was dissolved when B.M.C. took over, but it has retained the brand name E.C.C.O. since. Sales in South Africa continued through an agency in which B.M.C. had a 50 per cent equity holding. In the U.K., Allied Meat Importers (A.M.I.) was set up in partnership with a British firm of meat traders. B.M.C. took over A.M.I. and arranged a management contract with the U.K. partner, Meade Lonsdale. In 1977 the Commission bought a cold store which can now hold 7200 tons of chilled or frozen meat. There is also a blast freezer capable of freezing 50 tons in 24 hours. The whole operation, A.M.I. and E.C.C.O. Cold Store Ltd, is now managed by B.M.C. itself through Botswana Meat

Commission (U.K.) Holdings. Supplies from other sources are also handled so that, if supplies from Botswana are interrupted, B.M.C. (U.K.) continues to operate.

Foot-and-mouth disease is one of the major constraints to the development of meat export from Africa. European importing countries are acutely aware of the risks of importing exotic strains to which their cattle would be particularly susceptible. That Botswana has been able to overcome these fears is largely due to Dr Jack Falconer, Director of Veterinary Services. He built up a system of controls of livestock movements and annual vaccination of cattle in the northern part of the country, the most vulnerable to intrusion of the disease. The country is divided into zones separated by cordon fences (see Fig. 6.4). A permit is required to cross from one control zone to another and this can only be done at certain points where a quarantine is maintained.

The E.E.C. will not accept beef from animals which have been vaccinated during the previous 12 months. This also involves great organizational problems for the Commission. The country is divided into two areas, one supplying E.E.C. markets and the other supplying non-E.E.C. markets; kills must be segregated.

Fig. 6.4. Foot-and-mouth disease control fences, Botswana.

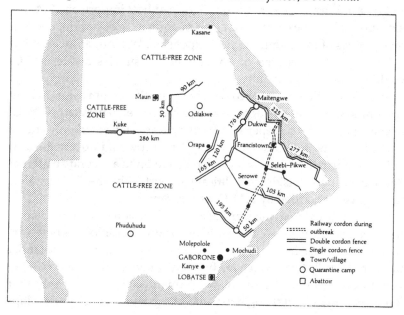

Evaluation

The Meat Commission is one of Botswana's major employers. Means are constantly sought to increase work opportunities, hence the preference for exporting boneless beef. The de-boning room takes 200 highly skilled men. In total B.M.C. employs 2000 people. Until the development of large-scale mining it was also the main source of income for the country and of export earnings. Operating in a complicated and extremely competitive market B.M.C. has consistently provided an attractive outlet for Botswana's livestock. Cattle raisers' incomes have benefited correspondingly.

Commercial viability has been preserved by the hiring of professional staff and the use of established marketing agents. Gradually B.M.C. took them over rather than attempting to create its own organization from scratch. In the factory, while Botswanans are being trained to take over all aspects of the running of the Commission, policy has been to recruit experts externally when judged necessary.

A vital factor in this success has been the relationship of B.M.C. with the Botswana Government. It has been left free to operate in a commercial fashion. Only very occasionally has pressure been applied in favour of political objectives which were commercially costly. The Commission has always received strong backing from the Government; its political support comes from cattle producers. Government lobbying in Europe after the E.E.C. stopped beef imports in 1974 was very successful. The construction and maintenance of cordon fences in remote areas of the country is very expensive. Without a Government willing to make this type of investment the success of the B.M.C. would have been impossible.

There were differences, of course, between the Commission and the Government. It took a lot of pressure, for instance, to persuade B.M.C. to control its effluent properly. A tannery was established only after it seemed that another Government corporation would do it instead. It took a decade before a decision was taken to build a second abattoir at Maun, a location considered politically important by the Government, but originally regarded by B.M.C. as too remote.

It can be argued that Botswana Meat Commission has served mainly the interests of the richer members of society. Since its business was livestock, this was inevitable. At one time it slaughtered a substantial number of small stock, which are frequently kept by those who are too poor to own cattle. The reason why this declined is probably that domestic demand following Botswana's very rapid economic development increased to the point that there was no longer a surplus for export.

Another possible criticism is of B.M.C.'s refusal to organize purchasing nearer to the tribal producers. Conversely, it can be argued that it was for the government to ensure that assembly was fair and efficient. The fact remains that the Botswana Meat Commission has been a highly professional undertaking and has been an outstanding model of livestock marketing in a difficult environment.

Cyprus Potato Marketing Board

Potatoes are the main export crop of Cyprus and an important source of foreign exchange. Of the spring crop harvested from early April through June 129 000 tons were exported in 1982. The winter crop planted in August, primarily for domestic consumption with exports amounting to 14 000 tons was of the total value in that year $38 million.

The Board was established in 1964 'for the regulation and control of exports of potatoes and their marketing and matters connected therewith'. It has an export monopoly. No producer can sell potatoes and no buyer can purchase potatoes for export except through the Board or with its permission.

The Board consists of the chairman and twelve members appointed by the Minister of Commerce and Industry. Out of the twelve members, six represent growers, two the exporters, one the Ministry of Commerce and Industry, one the Ministry of Agriculture and Natural Resources and two the Greek and Turkish Cypriot co-operatives respectively (now separated). The main decisions of the Board must be approved by the Minister of Commerce and Industry.

The objectives set out for the Board are:
(a) to secure for producers the most favourable and economical arrangements for the grading, packing, transporting, marketing, export, shipping and sale of potatoes;
(b) to secure, as far as is practicable, the enhancement of the financial returns to producers of potatoes on a long-term basis.
(c) to stabilize returns to producers of potatoes by reducing, as far as practicable, through the operation of a Stabilization Fund, the effects of short-term fluctuations in world market prices of potatoes;
(d) to regulate, as far as is practicable, the supply of potatoes in relation to effective demand and, should the need arise, to make.

within the limitations of the world market conditions then obtaining and of the Board's financial resources, the most equitable arrangements for the disposal of surplus potatoes.

The first action of the Board, in practice, is to control the allocation of seed potatoes. The whole quantity of seed potato used for the spring crop (approximately 10 000 tons) is imported from north-west Europe (U.K., Ireland, Holland). They are purchased by the growers' co-operative system, but the Board determines the quantity to be planted in each area and the time. Village committees then allocate this quantity to individual farmers according to their production over the past three years, with some weighting in favour of the smaller growers and those who have no other sources of income. Potato picking proceeds according to instructions from the Board transmitted through its personnel, the co-operatives and on the radio. The Board takes possession at the packing house which it has on rent from the co-operative. Here all potatoes are graded by size and appearance, subjected to phytosanitary inspection and packed in plastic or hessian bags, in cartons or in bruce boxes according to the market in view. Shipping is hired by the Board through commercial shipping agents offering space on a tender basis.

Three different sales systems are used according to the market. The main one is the U.K. which imports 75–80 per cent of the Cyprus potato production. In the U.K. sales are made to some 25 to 30 primary buyers. These are divided by area – South and North England, Scotland and Northern Ireland. Within each area they are the only distributors. All receive potatoes at a particular time at the same price. They sell either to smaller wholesalers supplying small-scale retailers or to buyers for the retail chains. Dividing the country by area is one of the innovations of the Board. Formerly the private exporters tended to unload most of their potatoes in the south and to compete for what they thought was the best market. The Board has found that it often does best in the north. Consistently small potatoes ('scrapers' for cooking as new potatoes) are supplied to the British market. These are packed in 25 kg cartons.

In continental Europe, particularly Germany, the Board sells on consignment through established wholesalers. The third market area is the Arab countries. For this area the Board invites tenders. In 1983 it had accepted an offer for 600 tons monthly from November to May with the condition that it would put no other supplies onto this market during that period. This market takes the larger potatoes. Early January they were on offer in the wholesale market at Riyadh for S.R. 100 per sack of 14 kg, i.e. 20 cents per kg. The retail price for small Cyprus potatoes in the U.K. at that time was $1.00 per kg.

The slogan used in sales material for Cyprus potatoes is 'Better looking better cooking'. The suggestion of red soil around Cyprus potatoes has become one of their sales features. The soil is rich in minerals and this is said to contribute to the quality of the potatoes. It also improves their appearance. There is a story of a rival supplier who bought red soil to put round his potatoes to try and raise their value.

C.P.M.B. spends about $200 000 per year on sales promotion in Britain. It is not yet convinced that expensive television advertising directed at consumers will pay off. At present, it focuses on the commercial press and radio programmes directed at retailers. Its goal is to persuade them to have Cyprus potatoes on sale. Consistently it has featured quality – also full weight. An extra kilo of potatoes is packed in each container to make sure that it meets the weight indicated when it reaches the retailer.

The success of the Cyprus Potato Marketing Board is attributed to three factors:

1. Very good management; it has had – since it came into existence – the same manager. He lives for the excitement of selling to the best advantage. He is always ready to take decisions and does not give up when he knows he is right. At the same time he has been in harmony with a Board of which the majority of the members have been growers.
2. The Board has always maintained an open sales policy, treating all its buyers in one country equally and never going back on its word. When it sold with a condition that no more stocks would be offered for a further ten days, it stuck to this. If any buyer complains that a load of potatoes is not up to standard they are immediately replaced.
3. Absolute control of the supply – the Board exploited to the full its power to adjust the supply of Cyprus potatoes on any particular market to maximize sales returns.

Andreas Savvides, the general manager, came from a farming background and studied agriculture. He went by scholarship to the university at Salonika, then to Oxford and by F.A.O. fellowship to the University of California at Davis. As head of agricultural economics at the Research Institute in Cyprus he had written a paper arguing the advantages of a marketing board for potatoes. He had no direct commercial experience, but at this stage the Board was going to use the existing exporters. He learned to sell by working with them. In the end he proved the best salesman of them all – because he liked it.

He says that the changeover in the years 1970–73 from selling through exporters responded to a trend in the U.K. to move away from speculative purchase by tenders to steady handling of supplies from a consistent source. This could be done on a lower margin.

The Board started direct selling in June of 1970. The exporters had tendered the very low price of $35–36 per ton which the Board did not want to accept. The Manager had sensed that the price would be better in July. He took the risk of holding potatoes off the market for a month and then sold them directly for $70 per ton. This lucky transaction, he says, made the point. It proved the advantage of controlling the time of sale of the whole supply.

In the ever-critical farming environment of Cyprus the Potato Marketing Board keeps a modest profile. Its total staff including ongoing employees at the packing plant numbers 24. In Nicosia, it is headquartered in a corner of the co-operative building. In London, it pays the Cyprus co-operative organization $60 000 per year for the part-time use of its building and assistance of its manager and staff – half of what it would cost to run an office of its own. The manager, however, commutes weekly between Cyprus and London during the selling season. A committee of the Board also spends considerable time in England checking the condition of their potatoes as they come off the ship and observing the state of the market. This helps greatly in maintaining grower confidence in the Board's decisions.

The capital and short-term financing requirements of this type of board are not very great. Table 6.10 presents its balance sheet for 1981 and 1982. Depreciation is calculated to write off the cost of fixed assets on a straight line basis over the expected useful life of the asset concerned. Annual rates used are:
- Packing house 10 per cent.
- Cold stores 10 per cent.
- Plant and machinery 15–20 per cent.
- Furniture and fittings 10–20 per cent.
- Motor vehicles 15 per cent.
- Fork lifts 20 per cent.

Freehold land and tools are not depreciated. Stocks of packing materials are stated at average purchase cost. Stocks of potatoes and potato seed are stated at contracted or realized sale prices.

The Board started with a loan of $24 000 from the government plus a commitment to guarantee commercial bank loans. In practice, this latter has hardly been used. The Board has been under pressure to spread its needs between the three main commercial banks getting a preferential rate on the understanding that it brings in its export returns through them.

Credit to farmers for seed, fertilizer and, if necessary, cash comes through the co-operative organization. The Board's role is to pay farmers' sale proceeds directly into their accounts with the co-operative banks. It pays them 75–80 per cent of the expected price at the time their potatoes are delivered to the packing plant. Final payment from primary buyers comes after about 40 days, but buyers are expected to advance 50 per cent at the time the sale is made. Thus the Board has only to finance a relatively small part of the value of the potatoes through transport and handling. From the final payment to the farmer it keeps back its own costs, and contributions to an operating reserve

Table 6.10. *Balance sheet: Cyprus Potato Marketing Board, 30 September.*

	1981	1982
	\$ thousand	
Fixed assets	809	650
Current assets		
Stocks	2 192	2 837
Debtors	490	571
Prepayments	174	21
Bank and cash balances	3 175	7 243
	6 031	10 672
Less current liabilities		
Creditors	1 128	3 970
Short-term loans	133	48
	1 261	4 018
Net current assets	4 770	6 654
Capital employed	5 979	7 304
General reserve	3 810	3 939
Price stabilization fund	1 768	3 365
	5 579	7 304

which is used to finance salaries, equipment and packaging supplies, and to a stabilization fund. Income and expenditure are shown in Table 6.11. Annual exports are around 160 000 tons.

In Cyprus the stabilization fund remains firmly in the Board's hands, normally invested in bonds and fixed-term deposits. It was drawn down to zero in 1983, however, following the exceptional low returns resulting from frost damage to the crop, and its subsequent late arrival on the market.

Impact on the growers

Before the establishment of the Board the growers faced prices fluctuating sharply from year to year and within a season. Various types of contracts were tried. However, growers finding the market price higher than the contract price would fail to deliver the contracted quantity. Exporters finding the market price lower than the pre-fixed one, used various excuses or legal technicalities to reduce the quantity

Table 6.11. *Income and expenditure account: Cyprus Potato Marketing Board, year ending 30 September.*

	1981	1982
	$ thousand	
Income		
Sales proceeds	54 414	69 184
Interest receivable	145	201
Sundries	—	52
	54 559	69 437
Expenditure		
Selling expenses	20 178	21 113
Provisions for bad debts	—	105
Packing house expenses	1 019	983
Administration	402	452
Interest payable	145	74
	21 744	22 727
Net income	32 815	46 710
Payments to growers	32 570	45 589
Surplus/(deficit)	245	1 121
Price stabilization account		
Balance at 1 October	1 523	1 768
Transfer of surplus	245	1 121
Government contribution	—	476
Balance at 30 September	1 768	3 365

purchased. Efforts to sell potatoes co-operatively had met with little success, mainly due to the structure and characteristics of supply.

Growers did not know whether or not they would dispose of their production. Unco-ordinated sales led to destructive competition between the exporters.

Following the establishment of the Board, potato production in Cyprus almost doubled from 1965 to 1970. From 1970 the seed allocation was gradually reduced from 14 000 to 9000 tons to keep up the price. The short supplies in Europe in 1975–77 provided an opportunity for production to be expanded again to record levels. However, in 1978, production and exports were cut by 30 per cent. This became necessary for marketing reasons and in compliance with a gentlemen's arrangement that Cyprus imports should not affect early growers in U.K.

In 1978, potato growers selling through the Board numbered 4296. Of these, 2425 delivered between one and 20 tons, 1273 delivered 21 to 50 tons, 423 over 50 and 175 over 100 tons each. A negative feature for some growers is the limitation placed by the Board on the quantities they may grow for export. This may inhibit expansion by some of the more efficient growers and obstruct the entry into production of newcomers.

The Board's margin is a very small part of the total marketing cost (see Table 6.12). Whether better returns to growers could be obtained under some other system, e.g. direct linkage by contract with major retail chains featuring Cyprus potatoes as sales leaders, or with more

Table 6.12. *Marketing margins for Cyprus potatoes sold in the U.K. – 1982.*

	Per cent of consumer price	
Consumer price		100
Retailer	29–32	
Wholesaler	7	
E.E.C. duty	11	
Handling, warehousing in U.K.	6	
Freight	9	
Packing, transport, administration	7	
Total marketing margin	69–72	
Producer price		31–28

intensive consumer promotion is debatable. The overall view of farmers in Cyprus is that the Board has been very successful. It has become a model for application to other kinds of export produce.

Attributes, advantages and support needs

Parastatal enterprises, marketing food and agricultural products, fall into three broad types according to the powers and responsibilities assigned to them by governments. These are:

1. *Parastatals stabilizing prices by trading alongside other enterprises* They operate in open domestic markets, buying when prices fall below some minimum level and selling when they go above some maximum. They do this by accumulating and releasing buffer stocks. Normally, imports and exports are controlled by the government in support of these stabilizing operations. Among the case studies, this type is illustrated by BULOG, the national foodgrain authority of Indonesia. It implements minimum prices guaranteed to farmers and ceiling prices to consumers.

2. *Export monopoly marketing enterprises* With a legal monopoly of the export of a particular product they can achieve greater bargaining power, apply uniform quality grades and specifications and adjust the quantities and types offered on particular markets in order to maximize total returns. They may also stabilize prices by maintaining reserve funds.

    The Cyprus Potato Marketing Board, the Botswana Meat Commission and the Zimbabwe Cotton Marketing Board have clear monopolies of export trading. The main objective in Cyprus was to control the quantities offered on particular markets by time and place. In Botswana the need to meet the animal disease transmission control and sanitary requirements of the intended markets was a primary consideration in centralizing live animal and meat exports under one official body. High confidence in its collaboration was essential if Botswana meat was to be allowed into European and other markets.

3. *Domestic monopoly marketing enterprises* These are the sole authorized traders (and processors) of designated commodities in specified areas or marketing channels. They can also separate and time market flows to maximize returns to producers and to stabilize supplies and prices. An additional consideration in the establishment of such monopolies can be to concentrate, under one manage-

ment, sufficient turnover of particular produce to justify investment in advanced processing and other facilities required for effective marketing. This type is illustrated by the Kenya Tea Development Authority (K.T.D.A.). It has control of tea planting in designated areas and a monopoly of sales from the growers there in order that it can maintain the quality of tea produced and so assure them of favourable prices. For the cotton growers of Zimbabwe, maintenance of high quality to maximize export returns was the goal in accepting detailed control of what they could plant and how it should be marketed.

In practice types 2 and 3 often overlap as illustrated by K.T.D.A. and the Zimbabwe Cotton Marketing Board. The Meat Commission in Botswana also sells on the domestic market, but not with a monopoly.

Attributes

In addition to the specific features set out above there is the more general attribute that these enterprises are established by government, are dependent on it for financing, and subject to its policy directives. Normally the Minister of Agriculture or his representative is chairman of the directing committee. There may be several other official representatives. It can be counted upon to reflect government policy.

In contrast, however, to direct government departments, a parastatal has considerable autonomy in day-to-day handling of funds, staff management and operational decision-making. An executive director or general manager is appointed. Within broad policy directives he can have, as demonstrated in the Cyprus case, considerable freedom of movement. Such a parastatal has, therefore, the capacity to take a fully commercial approach to marketing and perform in much the same way as a large private enterprise.

Many of the early marketing boards were set up at the request of producers, and farmers' representatives dominated the directing body. The role of the Minister's representatives was to guard the public interest. In the developing countries the pendulum has since swung the other way. Farmers' representatives have little say: dominant is the political interest of government.

An attribute of a government enterprise that appeals to many people is that it is not driven by the profit motive. The legislation setting up the Zimbabwe Cotton Marketing Board declares specifically that it is a non-profit-making organization. This can, however, be a very naïve consideration from the point of view of both producers and consumers. The operating margins of government marketing bodies often greatly exceed those of other enterprises because their costs are not under the pressure of competition and they have no direct incentive to seek out and adopt the most efficient methods of carrying out their marketing responsibilities. The interest of their staff can become an end in itself as distinct from those of the producers and consumers they are supposed to serve.

Advantages

The great advantage of the parastatal marketing enterprise for the government is that it can influence operating policy directly. For a planning ministry a parastatal marketing body is a clean instrument for plan implementation; there is then no need to work out a set of incentives to induce an independent marketing structure to follow a certain course.

In particular, the public marketing enterprise can be used to implement a guaranteed minimum price. This counts high for many farmers. They want to know in advance the price they can expect to receive for their produce before they commit themselves to the expenses of production, or maintenance and harvesting in the case of tree crops. Where funds are available from outside agriculture, an enterprise such as BULOG in Indonesia can also become a convenient means of promoting output of particular products by means of support prices and ensuring that these are obtainable within easy reach of most farmers. A parallel government policy objective for which such an instrument is extremely useful is that of cushioning the impact upon consumers of seasonally variable supplies of basic food.

For various developing countries, parastatal marketing structures were an advantage in obtaining aid finance. They provided a way round the reluctance of some aid agencies to appear to be helping particular private enterprises or persons. They also met the administrative requirement that there be a responsible national counterpart agency for a particular aid programme or loan, a requirement for which an array of private enterprises of varying size and stability would be less acceptable. Some of the crop authorities in Tanzania reflect World Bank interest in a counterpart agency that combined coverage of production and marketing.

Another advantage of a public enterprise monopoly is that it may be able to attract capital for investment in processing facilities which are needed, where private enterprise could not. Such a situation can arise where a new process, or level of hygiene, for example, can only be implemented at a level of throughput that is beyond the scope of any one enterprise in an existing structure of competing small traders. This is well illustrated by the Lobatse abattoir in Botswana with its access to Commonwealth Development Corporation (C.D.C.) and government funds.

Support requirements

The basic requirement for successful operation of a marketing parastatal is that the government should understand clearly: (1) the objectives of the enterprise, their implications for policy and procedure, and that decision-making by the responsible ministry and other official bodies should not be in conflict with these and, (2) the capital and income requirements of the enterprise, and that if it is required to undertake loss-making activities the government will provide compensatory revenues.

A board trying to stabilize prices in a free market will not be able to cover its full costs from purchasing and marketing operations. Necessarily, it forgoes the profit opportunities open to commercial enterprises operating alongside. Governments have also been slow to recognize the additional costs of holding food reserves and make provision to meet them.

General indicators for the efficient performance of parastatals are:
1. clear definition of the terms of reference and remuneration of the Chairman of the Board of Directors, of the Board Members and of the Managing Director, normally the Chief Executive;
2. clear public accountability, with the designation of an officer, not necessarily the Managing Director whose focus would be on policy and its implementation, who would be responsible for answering detailed questions on the financial management of the parastatal;
3. maintenance of an accounting system which provides up-to-date information on the costs and earnings of a parastatal's various activities, as well as the conventional accounting required of any public body;
4. some flexibility in the planning of expenditures and the handling of surpluses so that operations are not straight-jacketed by the need for advance approval of detailed budgets and incentives for efficiency are not blunted by requirements that all surpluses be returned to the Treasury at the end of a financial year;

5. minimization of political pressures for the appointment of staff
   on grounds other than their qualifications for the work envisaged and protection against such pressures from other sources.

A common criticism of public marketing bodies is that they are
overstaffed when compared with private enterprises doing similar
work. There is a tendency for staff allocations which are adequate in the
initial years to become inflated later. New staff are added, but for
various reasons – complicated procedures, or protection in high places
– fewer staff, are released.

Monopoly marketing boards and similar state enterprises offer many
potential advantages. They can use sales devices that will raise farmers'
returns 'through the use of price discrimination in separate markets,
reduce their risks by providing a guaranteed market, and protect
consumers by guaranteeing supplies at reasonable prices'. To governments, they constitute a valuable instrument for policy implementation. The benefits that accrue in practice will depend on the policies a
board pursues and the efficiency with which it operates.

An important factor in the success of the K.T.D.A. has been iden-
tified as a state of tension between the interest groups concerned – the
growers, buying centre and tea factory personnel, senior management,
the government, and outside partners in the financing of the Authority. Each group is watching the others and has an incentive to make the
Authority work efficiently. Against this may be contrasted the compla-
cency sometimes attributed to official bodies proceeding in a routine
bureaucratic fashion, protected by civil service conditions of employment and ministerial powers of self-defence. It follows that one important line of support a government can give to a state marketing
monopoly is to provide adequate means whereby users of its services
can make their comments and criticisms felt. These include:

1) insistence on timely provision of detailed annual reports, duly
   audited, and available for inspection by all interested persons;
2) provision for representations by committees of farmers, and other
   groups on a regular basis;
3) opening the press and other media to discussion of parastatal
   performance by university and other independent, informed persons;
4) periodic establishment by the government of committees of enquiry
   composed of qualified persons and empowered to take up information from farm, consumer, trade and other relevant sources.

Issues for discussion

1. Is there an official policy in your country regarding the role of government enterprises in food and agricultural marketing? If so, set out this policy and examine how far it is applied in practice.

2. Identify the parastatal marketing enterprises in your country. Under what circumstances were they established? Have there been any official reviews or enquiries regarding their performance? What were the main conclusions of such reviews?

3. If there is a public organization in your country, charged with stabilizing prices of a basic food grain to producers and consumers, prepare a profile of its operations. What are the policies it follows in establishing the prices at which it buys and sells? How does it make up any deficit on such operations? What subsidies does it receive in the form of capital provided without interest charges or not charged at market rates? Subsidies on transport costs? Exemptions from taxes? Which groups of producers and consumers benefit from its operations? Which are disadvantaged?

4. If there is a parastatal in your country with an export marketing monopoly how is this monopoly used? What are its main sales policies? What use is made of quality grades, exclusive contracts, timing of marketing offerings, use of market segmentation devices, brands and sales promotion? Assess, on the basis of available information, the advantages of its monopoly position as against alternative export structures, e.g. sales by open auction, through competing private exporters, through a transnational with established retail outlets and brands in consumer markets.

5. If there is a public enterprise with a monopoly of sales of certain products in designated areas within your country, review its operations in terms of the use made of this monopoly. What are the advantages and disadvantages? Which groups of farmers, consumers, traders benefit most? Which are disadvantaged? What are the direct costs of this monopoly to the economy, and the indirect costs?

6. Is there an example in your country of an enterprise assigned a monopoly in order to secure for it sufficient turnover to justify a major investment in processing or similar equipment and technology? How has it performed subsequently? Assess its efficiency by applying available criteria such as comparison with similar enterprises in the same or other countries, maintenance and/or expansion

of throughput and sales, returns to producers, foreign exchange
gained for the country, etc.

7. What official provision has been made for checking the performance
   of state marketing enterprises in your country? Do the reports
   issued provide sufficient relevant information? How should they be
   strengthened? How can farmers, consumers, other market participants voice their comments and suggestions?

8. On what basis are the managing directors, other managerial staff
   and other employees appointed to state marketing enterprises in
   your country? What are the incentives to them for efficient performance? What steps are taken to limit misuse of authority?

7

Developing an effective marketing structure

The focus on rural development and equity in the 1970s stimulated appraisal of marketing mechanisms not only for their ability to move produce to the best available markets with a minimum of waste and cost, but also for the consideration and assistance they gave to the smaller producers. Were they treated fairly at the local buying stage? Were they advised what to grow and when, and on how to raise its market value? Were they helped with credit, and with access to the inputs they needed?

The following structures have been used in providing marketing and associated services for small farmers in the developing countries.

- Independent private firms operating within some institutional framework such as assembly and auction markets or exchanges, possibly with some mechanism for cushioning extreme price fluctuations.
- Transnational companies bringing processing technology, economies of scale and established market outlets.
- Farmers' associations or co-operatives.
- Marketing board or other state agencies including special area and development authorities.

The relative advantages of these alternative forms of enterprise in a small farmer context are summarized in Table 7.1.

The most favourable position for the small farmer has often been that of participant in a production/marketing contract system for a particular crop organized by an enterprise with assured market outlets for the crop after processing. He receives a full set of services on credit. The quality of the extension assistance provided far exceeds that available normally because it is tailored to the needs of the market outlet served and the processes used. It is likely to be based on specific research and be backed up by the direct provision of seeds, pesticides, fertilizer on credit, together with day-to-day advice on how, and when, to carry out production operations and the harvesting and handling of the product.

Typically, this kind of marketing service has been offered by the transnational – British American Tobacco Co. for tobacco, Cadbury for cocoa in India, cotton companies in francophone Africa, for example.

Only a small proportion of farmers can obtain such contracts. For the great majority the first buyer of their produce is a private enterprise trader in their village or at a nearby market. In an effort to ensure that small farmers, in particular, received the services they needed, many governments embarked on integrated development programmes. With aid agency support they set up development authorities and multipurpose co-operatives linked to marketing boards. The intention was that models tried out on an area basis should be replicable over the country. The costs of maintaining these bodies in the face of declining government revenues and aid inflows since the recession of the early 1980s has brought a new realism to this policy area. The possible alternatives are reviewed in this light.

Table 7.1. *Relative advantages for small farmers of alternative marketing structures.*

Marketing Structures	Sales position of small farmers vis-à-vis larger farmers	Sales position of small farmer vis-à-vis buyer	Extension type assistance
Independent private firms	Bargaining weight can be used against small farmer	Advantage of access to alternative outlets	Advice based on local experience
Marketing and processing transnational	Equitable prices if can make a contract	Dependent but secure if can meet product quality requirements	Can be direct and intensive
Co-operatives	Equal if operates successfully	Favourable provided co-operative operates efficiently	Collaborate with government service
Marketing board/ state trading agency	Equal price if can reach official buying station	Depends on access to buying station; may be subject to illicit charges	Usually left to government service
Development company/ authority	Equal prices	Protected provided can meet product quality requirements	Can be direct and intensive

Indigenous private enterprise

Individual private marketing enterprises are well adapted to provide a range of positive contributions towards marketing efficiency and economic development. They have demonstrated themselves well suited to:

1. take advantage of and exploit unforeseen opportunities and follow up new ideas;
2. start up and go a long way with very little capital. Private marketing enterprises are great builders of capital assets. Their operators tend to be economical, even parsimonious, in their personal expenditure, very careful in their business outlays, stringent in their requirements of performance from paid staff;
3. operate at very low cost. Only those staff who make a positive contribution to the enterprise are employed. Full use is made of family labour available at no cost. Outlays on equipment and other

Seed/planting materials provision	Fertilizer supply	Other credit	Government support required
May supply on credit	May supply on credit	Consumption credit often available in addition	Provision of market infrastructure and information services; maintenance of competition, some price stabilization
Direct supply on credit	Direct supply on credit	No	Should negotiate participation for small farmers and prices
May arrange a supply	Direct supply on credit	May supply where supported by co-op. bank	Continuing financial support, supervision and protection generally needed
Assistance rare	Assistance rare	No	Insistence on measures to help small farmers at rural buying points
Direct supply on credit	Direct supply on credit	No	Major financial input or privileges usually required

capital expenditures are commonly kept to the minimum and
delayed until proved indispensable;

4. show ready initiative and quick response to changing situations
   (because decision-making is concentrated);
5. extend the marketing operation with high confidence and low risk
   using family ties and kinship linkages. Where the infrastructure for
   marketing is at an early stage of development, reliable means of
   communicating information, sales commitments and financial proceeds are important.

A continuing sanction against inefficiency in a private enterprise is
that, unless there are barriers to the entry of new firms, it will lose
customers and go out of business.

Areas of marketing where private enterprises tend to perform better
than others include:

1. *Perishable products* Variability in quality, a tendency to deteriorate
   quickly if not held in special storage or processed, and sharp
   changes in price in response to variable supply call for rapid
   responses on the part of enterprises marketing such products.
2. *Livestock and meat* The variability of the product, and the need for
   judgment in appraising quality and value and for care in handling
   to avoid losses give an edge to direct decision-making. The predominance of private enterprise in the marketing of livestock and meat
   also reflects a reluctance of many people to come close to the
   realities of this trade.
3. *Combined purchase of produce and sales of farm inputs and consumer goods* When the quantities supplied and taken by each
   customer are small and varying, considerable local knowledge,
   patience and willingness to work over a wide range of hours and
   locations is needed. Prices may have to be adjusted at each transaction, and complex small-scale credit arrangements maintained, if
   such an enterprise is to serve its clientele well. Often only a family
   enterprise, with a wife or child minding a shop while the husband
   goes out on rural purchasing and sales rounds, can provide this
   service economically.
4. *New and highly specialized activities in marketing* Characteristi-
   cally, these are the outcome of an individual initiative, not a
   planned development by a committee or a government department.
   Not all such initiatives are successful over the longer run. However,
   leaving no legal scope for private marketing enterprise can shut the
   door on the exploitation of unforeseen opportunities.

Transnationals

Potential contributions of the transnationals to marketing development and efficiency are as follows.

1. *Finance* Generally they are in a position to mobilize capital from the lowest cost sources. It can be brought in as equipment, improved seeds, strategic supplies and skilled management and technology for which foreign exchange would be needed in any event.

2. *Applied technology* Developing countries face the risk of selecting unsuitable designs and equipment and the problem of putting new plants into operation and maintaining them. Engaging an enterprise with demonstrated experience in applying a desired technology and in a position to keep it up to date is often the safest and, in the longer run, least expensive way of acquiring it.

3. *Management* When qualified management experienced in specific lines of marketing comes with a transnational, it is an immediate advantage. Local personnel can learn from it by working with it. The cost of maintaining expatriate managers will lead the transnational into promoting nationals into their place as soon as they are sufficiently competent.

4. *Quality standards and presentation* The transnational experienced in maintaining high standards can help a developing country meet such requirements for successful export marketing. It can also reduce quality risks to domestic consumers, and help adapt domestic agriculture to produce raw materials with the required attributes.

5. *Market access* In export sales a close link with an enterprise which has established outlets in major import markets is a great advantage. Experience shows that, when prices turn down, exporters with continuing distribution arrangements in the importing countries hold on to their market and the independents lose out.

6. *Bands* These carry great weight with consumers, and with the wholesalers and retailers who serve them. An agreement to sell through the owner of an established brand enables the producer to share in the benefits of past outlays on its promotion.

Market sectors adapted to transnational participation

Export marketing is the transnational field *par* excellence. Here, knowledge of the required technology, close familiarity with the import market's requirements, and an established position in it are of strategic importance. Processed products sold by brand have high priority, e.g. Del Monte canned pineapple and Nescafé soluble coffee. Also favoured are perishable products that can reach a distant market under integrated management and be sold by brand, e.g. bananas.

Farm supply marketing Economies of scale favour transnationals in the development and distribution of higher yielding seeds and poultry strains, and specialized livestock feed ingredients, pesticides, etc. Fertilizer, however, is now bought and sold in the developing countries by ingredient specification leaving little advantage to proprietary brands.

Domestic marketing Here the main opportunity for the transnational is in the application of advanced technology and associated commercial management and assurance of after-sales services. The introduction into Pakistan by Corn Products Inc. of a new technology for extracting starch products from maize led to a great expansion in its cultivation. Over-promotion by transnationals of branded consumer goods can nevertheless lead to a misallocation of consumer incomes, e.g. towards the consumption of baby food instead of mothers' milk. They may also involve a substantial import component.

Co-operatives

The marketing efficiency of a group of farmers is increased by their selling together where they can
- benefit from economies of scale in the use of transport and other services through increasing the volume of a commodity handled at one time;
- raise their bargaining power in sales transactions.

Conditions recognized as favouring co-operative marketing are:
- specialized producing areas distant from their major markets;
- concentration and specialization of production;
- homogeneity of production and output for market; and
- groups of farmers dependent on one, or a few, crops for their total income.

Factors favouring successful co-operative marketing are:
- availability of local leadership and management;
- a well-educated membership;
- a membership with strong kinship or religious ties.

On these bases some outstanding co-operative marketing operations have developed, such as the Ha-ee group in Korea.

More generally, co-operatives have shown themselves well suited to undertaking:

- assembly of fairly standard, not very perishable, products for sale on pre-established markets – where the price risk is small;
- distribution of a fairly standard, not very perishable, farm input such as fertilizer, also where pricing is pre-established.

Thus the assembly of coffee for export through a monopoly board has been a successful area of co-operative marketing in various African countries. Fertilizer distribution is a classic area of farmer co-operative activity in the developed countries. In the developing world, coffee, banana and other grower co-operative systems combine it conveniently with marketing the crop on which it is used. This constitutes a practicable basis for distribution of inputs on credit.

In general, co-operative systems have difficulty in matching the cost efficiency of competing private enterprises. There are grounds, nevertheless, for keeping them in existence as an alternative channel. They constitute a protection against the less competitive marketing structure that might prevail if they were not there. In many parts of the world the disappearance of a co-operative has been followed by less favourable terms from the private traders with which it competed. In India the co-operatives were allocated 40 per cent of the total supply of fertilizer for distribution, backed by the co-operative credit system. Co-operatives have been assigned local monopolies as assemblers of smallholder coffee or cotton. A protected role as handler for the government, or for a marketing board, of some standard product maintains a farmer co-operative coverage of the area concerned. This then constitutes a base from which an individual co-operative with strong leadership can undertake a range of other activities to help its members, as illustrated in the Taiwan case.

Parastatals

These are autonomous in day-to-day operations but directly responsive to government instructions. They are convenient vehicles for the application of public capital, implementation of government price policies, and assignment of marketing monopolies where these are judged advantageous.

1. *To moderate supply and price fluctuations* in food grains on domestic markets, a parastatal stabilisation agency can buy into, and sell from, a buffer stock. Most African, Asian and Latin American

countries have established such mechanisms to implement minimum prices to producers of major food grains and to protect consumers against prices likely to cause hardship. Sharp variations in price can be caused by marginal surpluses and deficits: buying into, and selling from, a buffer stock, some 5 to 15 per cent of the marketed supply of the product concerned is normally sufficient to eliminate wide price extremes. Confining the operation to such proportions limits the capital and subsidy required from the government; it leaves the bulk of the trade to free marketing enterprises, generally able to operate at lower cost because they have lower overheads and can select their transactions to match their resources and convenience.

2. *Export marketing monopolies* can obtain higher returns for growers if they control enough of the total volume going on to a particular market to be able to influence prices. Within its own seasonal niche on the U.K. market, the Cyprus Potato Marketing Board has done this. With 40 per cent of the export markets each for their long staple varieties, cotton export monopolies in Egypt and Sudan can manage them fairly well. The main import markets for cocoa were dominated by the Commonwealth West African export monopolies when they sold together, but this is no longer the case. In Zimbabwe the benefits from maintaining high-quality standards for specific buyers have been demonstrated. However, where buyer preferences vary, a monopoly board may obstruct price signals seeking to adjust production to their requirements.

3. *Monopolies in domestic marketing* are assigned to parastatals to concentrate sales of produce through a particular processing plant to justify the investment, to facilitate collection from small farmers of credit repayments and other dues, and to implement market separation programmes whereby higher overall prices can be obtained.

There are *commodity* marketing situations where parastatals are common, and others where for practical reasons they have been found less convenient and effective. Major food grains – maize, rice and wheat have priority. Less 'political' grains and pulses, including those often used by lower income consumer groups, receive less attention – because of governments' needs to limit activities that might call for eventual subsidization. Coffee, cocoa and cotton typically sold by standard quality specifications are widely handled by parastatals. Tea and tobacco requiring direct examination of samples are more often sold by open auction. Livestock and meat, perishable fruits and vegetables and relatively perishable tubers also tend to be left aside.

Constraints and support management

Private enterprise The classic concern about structure of private marketing enterprises is that they will collude to keep prices down to producers and up to consumers. The remedy is the entry of new enterprise. Appropriate government policy is to encourage the development of competing enterprises by alleviating obstacles to entry of new firms and facilitating access to information and to capital.

Attitudes towards private marketing enterprises are sometimes prejudiced by the fact that many of them are in alien hands. They work hard to establish a base for themselves and are free of the social and traditional restraints which bear on the performance of long-time residents. The best ploy for such countries is to learn from the aliens and absorb them. Where religious and other factors inhibit integration the aliens can still be obliged to take on local partners and form registered companies in which local interests have a specified minimum share.

In various developing countries, private marketing enterprises have been considered too many and too small. In an economy where small-scale producers and low-income consumers are also numerous, small marketing enterprises have a role that is strategic. They provide services that would otherwise not be available. Provided conditions are favourable, some private enterprises will develop on to a national scale.

An area of marketing may come to be dominated by one large private enterprise. If this is based on efficient use of scale economies, and the bulk of the profits is ploughed back into productive investments, this can be directly beneficial to an economy. If the investments are integrated through the marketing enterprise they are likely to lead to an effective economic development. In adverse contrast is private enterprise that retains a dominant position through corrupt politicians and 'mafia' style protection. The establishment in the 1960s of the Hong Kong Market Authority to market wholesale produce for producers on commission was a response to such a situation. With the balance corrected, direct transactions between private enterprise suppliers and buyers now proceed in parallel with the official market.

Transnationals The reservations of developing country govern-ments vis-à-vis the transnationals have focussed on the risks of becom-ing dependent on them, and so dominated by them. Illustrative was the concern over the withdrawal in 1984 of Gulf and Western Inc. from its sugar and other operations in the Dominican Republic. There is also fear that transnationals will use use their resources to influence local politicians and government decisions.

In fact, much of the steam has now gone out of the issue of transnational power.

1. Because of the uncertainties of foreign investment the transnation-als have tended to shift from production in a developing country to the sale of technology, management services and marketing.

2. The panorama of transnational is now much wider – no longer conspicuously U.S.A.- and European-based with its aura of neocolonialism. It includes many with their headquarters in other countries, e.g. Japan, and in some developing countries.

3. The form of transnational is becoming more varied and more flexible including banks, retailing firms, consulting firms and train-ing agencies.

4. Transnationals have learned to accommodate themselves more to the needs of the developing countries.

Transnationals have been shown to be very much subject to organized labour and political pressures. Alternatives are available to take their places – as was the case in the Dominican Republic. The situation has become one where the government of a developing country can assess the benefits that a transnational investment or collaboration can bring and bargain over the terms – witness Cadbury in India and Mumias Sugar.

Co-operatives Characteristic handicaps of marketing co-operatives are a) lack of own capital, and b) group decision-making. These follow directly from the democratic principle of the co-operative. If equal capital shares are to be subscribed by all members then they cannot be large, otherwise small farmers would be excluded. This means that most co-operatives depend on government finance for both fixed and working capital. Such capital tends to come on a standardized basis under decisions made at a distance. There is little commitment by the members themselves. Group decision-making implies less enterprise and ability in responding to changing market-ing opportunities. Directing committees of farmers often lack man-agement and marketing experience. This also reduces their willing-ness to offer an attractive salary and bonuses to a paid manager. They can be diverted from their long-run interests by influential local groups and politicians.

Stemming also from the co-operative principle is the need to maintain relatively complicated accounts. This can become a major preoccupation in environments where educational qualifications are limited, yet still not protect members against misdirection of funds.

Functions for which farmer co-operatives are well suited, situations favouring successful operation and grounds for maintaining them have been set out above. Promotion of co-operative marketing regardless of those parameters, particularly as a means of recuperating production credit, can involve governments in high support costs for the results achieved.

Parastatals Autonomy in day-to-day operations is vital for an enterprise engaged in practical marketing. While the autonomous parastatal is certainly better suited to marketing operations than a government department, many are still tied too closely to civil service salaries and conditions of employment. While ways are found to add staff, many parastatals find it extremely difficult to terminate them. Management capacity must be sufficient to overcome traditional attitudes and competing loyalties of staff, and the depredations of politicians.

The need for government to subsidize food grain stabilization parastatals can become a continuing burden. It can be kept down by maintaining a wider margin between buying and selling prices and applying price differentials for location, quality and storage. Local buying agents who already have an operating base and so incur lower overhead costs than direct purchasing stations can be used.

If a parastatal is given a monopoly, it is in a favourable position to avoid losses that must be met by government, but adequate checks on its efficiency are difficult to devise. In the absence of legal alternatives, producers and consumers will be obliged to use its services. It is on them that the burden of its costs will fall. Over the years 1971–79 the costs incurred by the monopoly board of Jamaica in marketing bananas in the U.K. averaged $100 per ton higher than those of bananas marketed in Germany from Ecuador. Along this latter channel, a national private enterprise was in competition with two transnationals. If a parastatal monopoly is maintained, there should be a clear technical justification – that it permits a certain marketing function to be carried out more efficiently than would be feasible otherwise.

Implications of various types of marketing parastatal, and of co-operative systems where they are given an official monopoly, are set out in Table 7.2.

Table 7.2. Implications of parastatal marketing enterprises.

Type of parastatal	Responsibility assigned	Nature of government commitment	Financial implications	Effect upon free market structure	Marketing and administrative skills required	Benefits for producers	Implications for domestic consumers	Conditions of implementation
Stabilizing domestic prices via buffer-stock on a free market	General control or trading monopoly of imports and exports, implementation of a pre-announced minimum price to farmers	Provision of initial capital; guarantee for commercial credit; support import/export controls	Unlikely to cover own costs so government must assign compensatory source of income or make up deficits	Free domestic trading continues subject to competition from parastatal	High ability to estimate supplies and prospects; judgment in buying into and selling from buffer-stock and skill in stock management	Announcement of guaranteed minimum price before planting facilitates production planning	Protected against very high prices by releases from buffer-stock	Product easily storable
Export monopoly trading	Sole exporter of designated products and implementation of pre-announced minimum price to farmers	Provision of initial capital, guarantee for commercial credit; support for export monopoly	Operating costs normally covered from margin allowed by price paid to farmers	Former exporters eliminated or required to act as agents for the Board	Can be operated with routine management skill; initiative and judgment needed to exploit monopoly weight and seize new opportunities	Announcement of guaranteed price facilitates planning but price may be low	None directly	Exports limited to a few points that can be policed
Domestic monopoly trading	Sole trader in designated products in defined domestic markets, implementation of minimum prices to producers	Provision of initial capital, guarantee for commercial credit; support for monopoly	May incur trading deficits if capital and operating costs excessive or is badly managed	Former traders eliminated or required to act as agents for the Board	Can be operated with routine management skill; initiative and judgment needed to seize new opportunities	Pre-announcement of a guaranteed price but it may be low	Price of products sold likely to be higher than with a free market	Feasible to control evasion of monopoly

In determining the most appropriate enterprise for a particular situation, local conditions can be decisive to an extent that is often glossed over. Where family allegiances are dominant, and the commercial infrastructure is uncertain, the more elaborate marketing organizations are handicapped.

Issues for discussion

Design suitable marketing systems for your country for a) a staple grain, b) a perishable product, c) an export commodity. If the systems chosen are different from those currently in place, how would you manage the necessary change from the present to the proposed system?

Further reading

Abbott, J. C. *et al.*, *Marketing improvement in the developing world*, F.A.O., Rome, 1984.

Abbott, J. C., *Agricultural marketing mechanisms; performance and limitations*, Staff paper, World Bank, Washington, 1984.

Artin, K. *et al.*, *Marketing boards in Tropical Africa*, Kegan Paul, 1984.

Asian Productivity organization, *Producers' associations for marketing farm products*, A.P.O., 4–14 Akasaka, 8-chome Minato-ku, Tokyo, 1982.

Casson, M., *Alternative to the multinational enterprise*, Moemolle, London, 1979.

C.O.P.A.C., *Commodity marketing through co-operatives – some experiences from Africa and Asia and some lessons for the future*, C.O.P.A.C., Secretariat, F.A.O., Rome, 1984.

Elgar, M., A failure in marketing technologies transfer: the case of rice distribution in the Ivory Coast *Journal of Macromarketing*, Spring 1980.

F.A.O., *Bibliography of food and agricultural marketing* F.A.O., Rome: issued periodically.

Goldberg R. A. and R. C. Mcginity, *Agribusiness management for developing countries – Southeast Asian corn system and American and Japanese trends affecting it.* Bollinger Publishing Co., 1979.

Goldberg, R. A. ed., *Research in domestic and international agribusiness management.* J.A.L., Press Greenwich Conn. 2 vols 1980 and 1981.

Harper, M. and R. Kavura, *The private marketing entrepreneur and rural development*, F.A.O., Rome, 1982.

Izraeli, D., D. N. Izraeli and F. Meissner, *Agricultural marketing for developing countries*, Vols 1 and 2, Halsted Press, J. Wiley & Sons, New York, 1976.

Kaynek, E. E. ed., *World food marketing systems*, Butterworth, London, 1986.

Killick, T., *Development economics in action*, St Martin Press, New York, 1978.

Lipton, M. ed., *Rural poverty and agribusiness*, Institute of Development Studies, University of Sussex, 1977.

Livingstone, I. and H. W. Ord, *Agricultural economics for tropical Africa*, Heinemann, New York, 1981.

Rao, T. V. and T. K. Moulik ed., *Identification and selection of small scale entrepreneurs*, Indian Institute of Management, Ahmedabad, 1979.

Van der Laan, H. L., *The Lebanese traders in Sierra Leone*, Mouton, The Hague, 1975.

8
Managing a marketing enterprise

The broad principles of management hold good for agricultural marketing enterprises in the developing countries as for other business enterprises. Adaptation will be needed to take account of varying conditions of resource availability and infrastructure. These principles also hold good for different types of marketing enterprises – whether individually owned and operated, transnational, co-operative or parastatal.

The approach followed in this chapter will be to introduce the main principles of marketing management and point to illustrations and discussion in the case studies. The great advantage of cases from real-life experience is to bridge the gap between statements of principle and precept and application in practice.

The cases also provide information on the application of marketing management techniques to various products and activity combinations. For a more systematic coverage of the marketing of agricultural products and inputs in the developing countries reference should be made to the series of Marketing Guides published by F.A.O., Rome and other specialized texts.

Planning operations

The essential starting point for any marketing enterprise is a clear idea of the market to be served and a plan of action for serving it. Kallu in Sierra Leone knew the market for his own produce. When he was drawn into acting as a wholesaler for his neighbours he did not appreciate the implications of their irregular supply. He then found himself worse off because he was doing more work yet receiving a lower net income.

A reconnaissance visit to appraise the market, ascertain conditions of entry and identify potential buyers is recommended for all new marketing operations. With the results of such an enquiry in hand, a marketing manager is in a position to acquire, select and present produce to meet these requirements, and to formulate a sales plan. Mary Jane of Dominica recognized that there was a surplus of agricultural produce on her own island. She had heard that prices were much higher on Guadeloupe. That she did so well was lucky, since we are not informed that she had made a prior investigation.

Market research

The larger the marketing enterprise, and the more it has at stake in terms of owner or client commitment, the more important it is that systematic research on the markets open to it be undertaken as a guide to policy. Market research can serve both to plan a new sales programme and to maintain or develop an existing market share. It can serve the interests of private firms, co-operatives, parastatals and those of agricultural regions or countries as a whole.

A research programme for agricultural products, for accessible domestic or export markets, should include:

a) official regulations and market preferences relating to sanitary requirements, quality standards, containers, inspection, etc.;

b) quota regulations and levies affecting volume, cost, and timing of sales in particular markets;

c) supply and price trends, overall and for different qualities and forms of presentation, by year and by season, also covering possible substitutes;

d) sales methods and agencies, and their respective cost;

e) prospective consumer demand, the share that could be obtained, and ways of increasing this.

Enquiry along these lines is designed to answer the following questions: What is the size of the market? What are the predominant consumer preferences? What is the scope for market segmentation, i.e. for separating consumer groups likely to pay different prices for basically the same product? What are potential bases for such segmentation – quality, packaging, branding, form of presentation, type of retail outlet? Is demand seasonally limited? How would it respond to measures to extend product availability by storage, selection of varieties, modification of planting time? Is demand likely to grow? What are the determining factors? Is demand seasonal or year-round? What financing would be required to cover shipping costs, delays in obtaining settlement, sales promotion, etc.?

Consumer preference surveys are used increasingly as a guide in the presentation and processing of produce, or to reveal reasons why sales of long-established items are falling off. The services of independent agencies specialized in this field can be used to advantage in developing new markets.

Determination by market research of the buyer structure can provide valuable guidance for policy on sales methods, packaging and presentation. In the U.K. where grapefruit are eaten widely at breakfast, the Israel Citrus Board sells them at a low price unwrapped simply stamping the word Jaffa on the skin. In Italy, where grapefruit are appreciated mainly by upper income consumer groups who have travelled, they are sold in units of three, with a brand on a plastic wrapper – at a price double that in the U.K.

Private enterprise and transnational marketing managers are generally free to adapt their operations to match market demands. Managers of farmer co-operatives and parastatals may be subject to specific constraints such as a commitment to buy all the produce that is offered to them. This can mean that secondary outlets have to be found for produce that does not match the requirements of the most favourable markets, or is in excess of the demand of certain large buyers who would accept supply contracts for specific quantities. It is still open to them, however, through the prices they pay to growers, to direct production towards market requirements.

Access to capital

If a marketing enterprise is to buy produce, transport or store it and resell, it must have sufficient funds to cover payment for the produce, transport, storage and handling costs until the proceeds of sale come in. This can be a matter of weeks as in the case of Mary Jane. It can be much longer if produce purchased at the harvest season is held for sale to consumers later in the year. Overhead costs of facilities, equipment and staff must also be met.

Most private marketing enterprises begin with some own capital, however small, add to this by borrowing from relatives, and then go to a bank or other credit source for a short-term loan to make up what is needed to begin operations. If the enterprise is successful and can provide adequate collateral such as a house, land, or business property, it can expect further bank finance in subsequent years, possibly on easier terms. For the future market kings of Sudan, the take-off point was when they had accumulated enough of their own capital by parsimonious living, opportunism and good fortune to become really creditworthy with the commercial banks. Soex, the bean exporter in Senegal, found his operations restricted by lack of capital. The commercial banks saw his enterprise as too risky to justify larger advances.

Selection and management of staff

Working partners who can be trusted are crucial in the development of a marketing enterprise in its early stages. Often it is a husband and wife team, a father with working sons, or partners who are already friends of mutual confidence that constitute the operational nucleus. Small private enterprises then take on additional labour as and when they need it.

The terms on which continuing staff are engaged should provide a clear incentive for performance. Salaries and prospects should be attractive so that employees are keen to hold on to their jobs: otherwise they should be paid by results. An employee will demonstrate strong evidence of self-motivation if:

1. he is properly trained in the work and knows what to do;
2. the desirable results of whatever he is asked to do have been defined;
3. guidelines or limits in terms of policy, expenditure, and time have been established;
4. he is left alone to do the job;
5. he knows he can go to his boss at any time for guidance or support when he reaches an impasse;
6. he knows he will not be berated if things do not work out exactly as the boss wanted;
7. he is immediately praised for the things he does well.

Al Haji, the marketing board assembly agent in North Cameroun paid his buyers 24 cents per sack purchased. He provided them with a motorbike to scour the countryside. Clearly, he was aware of the importance of delegating responsibility. By defining the requirements of the job and the rules to be followed, offering a recompense directly related to achievement and furnishing the means for it, he mobilized their natural abilities, local knowledge and experience.

Choice of site

Desirable attributes in a site for a marketing business are: convenient access to supplies of the produce to be handled, access to a good market for it, directly or via reliable transport means and agents, and prospects of growth. Bulungu was advantageous for Mafandala in Zaire because it was a critical access point for transport to Kinshasa, his main market centre and source of consumer goods. Matlhaku wisely opened a shop in Gaborone when he heard that it would become a capital city. In Illushi, Enebor was located in an area where rice production was expanding, but his business was handicapped by poor infrastructure. Surfacing of the highway and provision of electricity and piped water would be a great boon to him; but would it come during his lifetime?

Business management
Pricing

The initiative that a marketing manager can take in setting prices for the products he sells is determined by his market position. This can range from that of an irregular seller of small lots on a public market to a monopolist who feeds products on to the market at a rate designed to maintain a target price.

Illustrating the first extreme is the peasant woman of New Guinea (T. Scarlett Epstein, *Urban Food Marketing and Third World Rural Development*, Croom Helm, London 1982). She sets off to market with a mixed lot of vegetables and no clear idea as to what they are worth. She asks about prices from people she meets on the way. She starts to sell at the price most often mentioned. If there are no buyers she reduces the price. If she seems likely to dispose of her produce before midday she raises it. In the afternoon, with the prospect of carrying some produce home or giving it away, she cuts her price. Step-by-step she seeks the best set of prices consistent with disposal of her stock in hand.

The monopoly seller sets his sales price to maximize profits. He can supplement his personal knowledge of the market with in-depth studies to estimate the elasticity of consumer demand with respect to price. This information would tell him how much more consumers would buy if the price were reduced, and how far sales would drop if the price were raised. Using estimates based on these studies, he can then decide what quantity to put on the market in order to maximize returns to his enterprise and the prices at which this would be achieved. Thus the manager of the Cyprus Potato Marketing Board can tell his panel of buyers in England that he will sell them 50 tons each at a price of $40.00 per ton. He will allow them two weeks in which to sell this supply before he puts more supplies on the market. They are likely to accept this because they have his protection against a competitor selling the same kind of potatoes acquired at a lower price.

Between these two extremes there is a range of intermediate positions. Enterprises offering quantities too small to affect the market in which they sell must generally follow it. They can, however, try to sell part of their supply to special customers at a premium on the basis of freshness or some other attribute of quality or convenience. This is the position of farmers who sell part of their output retail, with the rest going to a wholesaler at whatever price it will fetch.

For many enterprises, the published price on some established wholesale market will be their main guide in pricing. Competition between buyers will have determined a margin reflecting transport, handling and transaction costs between the price at this market and at their own location. Sales proceed on the basis of the central market price less this margin (made up of these costs). Thus in Paraguay in 1983 farmers could make advance contracts with soybean wholesalers on the basis of a margin of $80.00 per ton below the price on the Chicago Exchange, even though their soybeans went mainly to Rotterdam. Some farmer co-operatives offered their members a choice of the current price at Chicago or the quoted *forward price* for the following month, or for the month after that, always less the margin.

Parastatal exporters of standard products such as cocoa, coffee and groundnuts tend to follow the established international markets. To enable them to clear their warehouses for incoming supplies, and reduce their bank borrowings, they sell a large part of their total supply at the current spot price. The rest is sold forward over the months when they can expect conveniently to ship it. For such products the spot and forward prices on the London and New York markets reflect the best expert opinion.

Purchasing

Marketing begins with the production or acquisition of products for sale. In planning production or purchasing, the marketing manager will take the following into account.

1. *Characteristics of the product* Does he know it sufficiently well to assess its quality and value? Can he pack, handle, transport and hold it for future sale without incurring substantial waste and deterioration?

2. *Ability to finance* Does he have sufficient of his own capital available or will he have to go to a bank for credit? Will he be able to get enough credit and at what cost?

3. **Price likely to be obtained on sale** Is this a sure price? If it depends
   on free market determination, is the current trend up or down?
4. **Prospective profit** Most marketing managers incorporate this into a
   target *operating margin* which they use in deciding when, and at
   what price, to buy.

This *operating margin* is made up of

- Direct costs — payments for handling, transport, market
  charges, sales, etc.
- Overhead costs — office expenses, salaries and social contributions for continuing staff, bank
  charges, depreciation of equipment and
  facilities. These are estimated per unit of
  quantity on the basis of recent records.
- Remuneration for — the net income of the operator or his enter-
  management and risk prise.

For convenience, marketing enterprises adopt a standard mark up for
their operating margin. This may be 5 to 10 per cent of the purchase
price for a wholesaler selling large lots fairly quickly; 25 to 30 per cent
for a retailer of perishable produce taking longer to sell, dealing in small
lots and carrying the risk of physical wastage and produce remaining
unsold; 18 per cent for a supermarket handling a mix of products
including many consumer essentials with ample shelf-life.

So a manager will buy produce that he thinks he can handle, that he
can finance, and that he believes he can sell with an adequate mark-up.
This will be his standard analysis. If he foresees some risk over the
resale price he may reduce the quantity purchased; or he may decide to
carry the risk in order to please a regular supplier and to be sure of
satisfying regular buyers.

For processing enterprises, particularly, it is convenient to secure
supplies via advance contracts with producers. See the Tabasco case
and Jamaica Broilers. This enables the processor to specify in advance,
variety, quality, maturity and standards for his raw material, and time
of delivery. He can then operate his plant more efficiently and be sure
of satisfying his customers' requirements.

Sales

Selling is at the heart of marketing. The owner of a small private
marketing enterprise is likely to undertake sales directly. He will draw
on his knowledge of the products in which he deals, of the market and
of his experience of human attitudes in bargaining. Illustration of

various pragmatic approaches to sales can be found among the private enterprise cases.

The owner would first delegate responsibility for sales decisions to family members that he could trust. In the Ogbomosho Co-operative Society in Nigeria several of the founding members shared responsibility for sales.

Establishment of a sales department headed by a sales manager becomes advisable as an enterprise increases in size and complexity of operations. Selection of a suitable person for this key post is crucial. He should have experience of selling in the markets envisaged: organizing ability and imagination will also be important. He should be open to new ideas and on the lookout for new opportunities.

Appointment of agents to sell for a firm at distant locations is a further phase of delegation and specialization. They may want exclusive rights in a defined market area as well as an attractive margin or commission if they are to put their full weight into selling for the firm. The system developed by the Ha-ee village group, whereby a group of wholesalers in markets in the same big city are kept under continuing pressure in selling for it, reflects a very strong market position.

Density of demand determines the number of distributors sought for Cyprus potatoes. A panel of 26 was used in the U.K. Over the whole Arab market one buyer was given exclusive rights for a six-monthly period.

Sales methods should take careful account of buyers' requirements. The Kenya Tea Development Authority offers tea at daily auctions so that its clients can inspect it before purchase. In contrast, cocoa and coffee are sold by description.

Ability to offer short-term credit can be very effective in facilitating sales. How far an enterprise should go in this direction depends on its cash flow position and its competition. Many wholesalers finance meat and vegetable retailers for a few days: 30 days' credit is common for more durable products, 60 days for fertilizer exports. In a strong sales position, Cadbury in India could require that its wholesale distributors pay cash for their supplies.

Advertising

How best to promote sales and how much to spend on this are continuing issues for the marketing manager. It has always been difficult to assess response to advertising; yet few marketing firms have felt they could do without it.

Some promotion is essential to make potential buyers aware of an enterprise and of what it sells. For products that can be branded a much larger outlay is justified. Differentiated by brand promotion from others basically similar they can be set in a continuing higher priced category. Great here are the returns to scale. A heavy outlay on advertising can be afforded because the unit cost is low if the volume of sales is large. In Turkey, Unilever-Is spent about 1.5 per cent of sales on advertising. (See the case study for details of its allocation between media and for designs appealing to family use of its products.)

Consumer-directed advertising can be very effective and also very expensive. For large-volume brands such as Jaffa oranges and Chiquita bananas it has paid off. On a national scale, 'Best Dressed Chicken' has done well for Jamaica Broilers. The manager of the Cyprus Potato Marketing Board preferred, however, to advertise in U.K. trade journals and maintain personal contact with his buyers. A spot on national television in the U.K. in the early 1980s cost over $1000 per second.

Co-ordination of enterprise promotional efforts with other related agencies and services is important. Thus mobile promotional vans sponsored by a fertilizer distributor should operate in conjunction with government extension staff in organizing field days and on-farm demonstrations. They should co-arrange that retailers in the area concerned have stocks conveniently ready at promotional prices together with explanatory materials and point-of-sale announcements.

Packaging

The art of packaging is to combine protection of the product with a presentation that helps to sell it, at an acceptable cost. In developing country markets where consumers are accustomed to accepting produce loose in their own containers and are primarily concerned to buy at the lowest unit price, expenditure on elaborate packaging would be wasted.

In sharp contrast is the outlay incurred by Soex on containers for green beans for transport to Europe by air. Light-weight protection of the product, and appeal on wholesale markets in Europe was critical for this operation.

Transport

Timely movement of produce from where it is in surplus to where it is wanted is at the heart of marketing. The main concern of the manager in developing countries is to secure a reliable low-cost service with maintenance readily available. To justify investment in a self-owned motor truck a marketing programme that will maximize its use should be in clear view. There should be pay loads in both directions. Mafandala's trucks brought beer and consumer goods from Kinshasa to Bulungu. When they went out into the country from Bulungu they carried consumer goods and brought back agricultural produce which was then made up into full loads for the longer journey to Kinshasa.

Refrigerated vehicles offer technical advantages in marketing perishables. They are expensive, however, and the risk of losses due to breakdown is high. Until it is clear that their use is justified by the marketing advantage, other ways can be used to control product deterioration in transport. For many years Unilever-Is distributed margarine in Turkey covered with mats soaked in water in ordinary trucks. Evaporation of the water had a cooling effect. Drivers were instructed always to park a loaded vehicle in the shade. The main supplier of meat from the Sierra to Guayaquil on the coast of Ecuador in the 1980s used an insulated van travelling by night, so avoiding the capital outlay on a refrigerated truck.

Arrangements for shipping produce by sea are illustrated in the case on cassava exports from Thailand.

Storage

How long to hold produce during marketing depends on the increase in price that will be obtained by selling it later, and the cost of holding it. Storage costs are made up of three main components:

- the rent of the storage facility;
- physical losses and deterioration of the product while in storage;
- interest on the capital represented by the value of the produce in storage.

The inter-relationship of these three factors is illustrated in the following example of the storage of apples under refrigeration in a near-eastern country (Table 8.1.) Thus while the difference in price between the harvest and low supply season was $60.00 per ton, the profit margin in storing until that time was only $4.70 per ton.

Losses on grain can be brought down to very low levels by drying before storage, use of insecticides, etc. Nevertheless, for storage of grain over eight or nine months to be economic a seasonal price increase of 17 to 20 per cent on the initial value is generally needed, depending on the conditions and interest rates applicable.

Al Haji used his house as a store. Establishment of specialized

Table 8.1. *Illustrative costs and margins in the storage of apples.*

	$ per ton
Estimated increase in price over six months	60.00
Storage costs	
Rent (or overhead cost for owned facility)	21.00
Storage losses (weight and quality)	23.80
Rent of crates	4.30
Interest (10%) on initial value of stocks	6.20
Total costs	55.30
Net margin	4.70

storage is a long-term investment project calling for a careful demand assessment. Convenience in location is all-important. Proximity to a mill has the advantage that the mill can be supplied as needed, with no further expenditure on transport. Convenient also for storage are places where a change of transport is obligatory, e.g. at a port or where feeder roads link with a rail line serving a distant market. Parastatals concentrating storage at central points sometimes forget that this can mean transporting grain back in the direction from which it came earlier in the year.

Ability to offer credit on produce in store is a great advantage. For this a bank is a good partner, also in financing the long-term investment in a storage structure. The two major grain stores in Ecuador undertaking storage for farmer or traders on a fee basis are both run by companies in which commercial banks are partners.

Extending the range of activities

From time to time most managers see convenient opportunities to add to their marketing business. These can be supplementing an existing marketing line with complementary products, undertaking additional functions, e.g. processing or retailing, strengthening existing operations by setting up branches or agencies to sell in new domestic or export markets. The decision should be based on the marginal return: that is to say profits on the new activity, after covering all costs directly attributable to it, are calculated as a contribution to meeting the general overhead costs of the enterprise, not after a proportion of these overheads have been allocated to the new activity. Use of a work sheet as illustrated in Table 8.2 can be helpful in making estimates. The

Table 8.2. *Estimating the advantages of taking on a new activity.*

	Year 1	Year 2	Year 3	Year 4	Year 5
New activity					
Gross additional income	....	....	....	....	....
Less Variable costs					
Labour	....	....	....	....	....
Transport	....	....	....	....	....
Sales	....	....	....	....	....
Sub-total	....	....	....	....	....
Less Fixed costs relating to activity	....	....	....	....	....
Financing of new equipment	....	....	....	....	....
Depreciation of new equipment	....	....	....	....	....
Continuing salaries	....	....	....	....	....
Promotional outlays	....	....	....	....	....
Sub-total	....	....	....	....	....
Total costs of new activity	....	....	....	....	....
Net contribution to general overheads	....	....	....	....	....

contribution of a new activity may be small or even negative in the first years; by the third or fourth year it should increase substantially if it is well suited to the enterprise and within the capacity of the personnel available.

Scope for additional activities in conjunction with agricultural marketing is considerable because of the seasonal concentration of many of the major operations and the need to maintain purchasing and supply services in rural areas with a low density of demand.

Business controls

Maintaining accounts

While enterprises extending over several countries have been built up by men who were illiterate, it helps to keep some accounts. Kallu found this out to his cost. At a minimum, a marketing manager should know how much he owns in relation to what he owes, and what he is earning in relation to what he spends. For these purposes he should have prepared annually a balance sheet showing his assets and his liabilities at a particular time, and an income and expenditure statement setting out clearly whether he has made a profit or a loss over a defined period.

Family enterprises will find such accounts useful as a guide to the progress of their business and to possible financial dangers ahead. They will be needed in approaching a bank for credit and making statements of income to tax authorities. Enterprises that are partnerships, joint stock companies, co-operatives and parastatals are generally under obligation to produce such accounts for shareholders, co-operative members and responsible departments of government. Balance sheets for marketing enterprises are presented as Tables 3.6, 4.4, 5.7, 6.4 and 6.10. Figures for the preceding year are normally provided to facilitate comparison and evaluation of changes between years.

Adequate provision for depreciation of fixed assets is essential if a financial balance is to be realistic. Depreciation allowances for various types of assets are illustrated in the Cyprus Potato Marketing Board study.

Useful indicators of the financial state of an enterprise are its net worth, i.e. owner's capital after deduction of indebtedness, and the current ratio, which is

$$\frac{\text{current assets}}{\text{current liabilities}}.$$

A current ratio of 1.6 to 2.0 is generally acceptable. This is an indicator of adequate working capital, though trading enterprises can work with lower ratios provided the level of debtors is low, stocks are well controlled, turnover is rapid and prices cover short-term financing costs. Some of these considerations are reviewed in the case of the West Cameroun Co-operative Union.

Tables 3.1, 3.3, 3.4, 3.5, 4.7, 4.9, 5.3, 5.6, 6.5, 6.9 and 6.11 are income and expenditure statements showing the net profit of an enterprise for a year. Comparison of the component figures with previous years can help explain why profits have declined or increased. To bring more precision into operating plans, managers can be asked to prepare income and expenditure budgets for a coming year. If the actual figures differ markedly from those budgeted an explanation should be sought.

One critical ratio is the rate of earnings on the capital invested. If, for some time, this is lower than that available elsewhere the owners of the capital will be dissatisfied and be inclined to withdraw it.

A strong cash flow position gives a manager scope for initiative and movement. A commonly used indicator of this is the ratio

$$\frac{\text{net income to shareholders} + \text{depreciation allowances} + \text{other income net of taxes}}{\text{capital expenditures} + \text{changes in inventories} + \text{dividend or interest commitments on capital}}$$

A ratio of more than 1.0 indicates a strong cash flow and high capacity to take on new activities and make new investments.

Keep watch on costs

The low operating costs of many family marketing enterprises stem from the direct relationship between outlays on services and the owners' pocket. The more use he can make of family labour and of own resources such as his house, outbuildings and a vehicle perhaps, the more he keeps for himself.

Under the pressure of handling a seasonal product, however, a larger-scale operator will have to engage whatever services are necessary to carry out the operation. Preferably such needs should be foreseen in advance, and the costs of using alternative suppliers appraised against their quality, reliability and timeliness. When the marketing season is over, the manager can do his cost accounting, setting out the supplies and services purchased and analysing their impact on the profitability of his operation. In Table 3.2 the various costs involved in getting green beans from farms on to a plane at Dakar airport are set out in dollars per kilo. In Table 4.7, major costs incurred by Unilever-Is in Turkey are compared as a percentage of sales. Such analyses bring out the relative magnitude of various costs and changes between years, so pointing to those meriting special attention in a subsequent season.

This type of cost control is just as relevant to parastatal marketing operations as to private firms and co-operatives. Comparison of the current cost of standard operations per ton of produce handled with those of previous years is one of the efficiency measures available to a monopoly enterprise. Thus the ocean freight loss has been a strategic indicator for the Sierra Leone export marketing board. A rise above the normal meant pilferage of cargo during port loading. A rise in the average turn-around time of vessels in port signified poor organization of transport or handling equipment and crews. A rise in administration and other overhead costs per ton could reflect reduced handlings because of a poor harvest; otherwise it would point to a need to simplify procedures, combine personnel responsibilities and reduce total staff employed. To permit such analyses, expenditures on major operating functions – transport, packaging, storage, port handling – should be accounted for separately, not lumped together under such headings as labour, services, etc.

Currency risks

Sudden changes in exchange rates can affect dramatically a marketing enterprise engaged in the export of agricultural products or distribution of supplies that are imported. If a change in rates is anticipated, the operator will do well to budget his outlays on replacements at the rate foreseen for the future.

High rates of inflation are common in many countries. The manager of a marketing enterprise must be aware of their implications. Selling on the basis of a 10 per cent margin to cover costs and profit will leave him in deficit if the currency has devalued 10 per cent in the meantime. To maintain his income in real terms his margin must be costs plus target income plus an allowance to cover the expected degree of inflation! An enterprise distributing fertilizer in Brazil in 1983 added a 10 per cent margin to its procurement prices to cover its own costs and profit, then a further 50 per cent to cover expected inflation. Depreciation allowances should also be based on the cost of replacement at current prices, not past costs of purchase or construction.

Insurance

Enterprises engaged in the marketing of agricultural products face a wide spectrum of risk. It ranges from price changes in response to events fully external to the enterprise to losses of produce in storage or transit and misuse of funds by responsible officers.

Protection against losses on forward contracts or stocks due to price changes can be obtained by hedging on a future market where one is available for the commodity concerned. In many developing countries access to such markets is limited by government policy and foreign exchange controls. However, sales agreements can be related to prices reported on recognized forward markets, as illustrated for soybeans in Paraguay.

Protection against many other risks can be obtained via insurance on a national basis. A co-operative can insure against the disappearance of a manager with its funds by requiring that he take out a bond for $5000 or $50 000 according to an estimate of the amounts at risk. A firm undertaking storage of produce for other enterprises can be required to obtain insurance against loss due to fire, its own negligence, etc. Insurance can be obtained against losses of produce shipped on consignment due to delays in transit and mishandling by the transporter or consignee. Where the risks of accidents, illness, etc. sustained by employees or users of marketing enterprise facilities are substantial, insurance against them should be obtained annually as a matter of course.

Use of computers

Computer technology opens the way to convenient access to data, speedy processing and analysis of accounts, and easy correction, reproduction and issuance of standard typed materials. Common uses of computers by marketing enterprises and support services are to:

a) maintain and provide easy access to inventories and expected arrivals at a number of depots or branches. Thus pressing a button will present on a screen an up-to-date summary of stocks at all its depots for the Director General of BULOG;

b) analyse sales records, programme sales forward and provide financial management information that will help marketing managers to maintain efficiency;

c) maintain records of salaries and social contributions of enterprise staff, prepare salary cheques, and make tax and other standard deductions where applicable;

d) prepare and address standard letters to potential customers;

e) maintain price and other information for a large number of markets and make it available on request;

f) calculate optimum solutions to problems involving simultaneous equations of input and price variables, and present alternative supply options. Thus least cost combinations of various feed ingredients to meet a required nutritional formula can be determined. Use of cassava chips for livestock feed in Europe has been favoured by access to this technology. Lowest-cost locations of fertilizer distribution depots can also be determined on the basis of demand patterns and transport costs.

Computer use commonly has three phases:

- programming the computer to carry out existing tasks more efficiently;
- finding new ways of increasing its effectiveness in carrying out these tasks;
- devising, on the basis of experience, new tasks for the computer that might not have been foreseen initially.

Effective use of computers is feasible where:

- electricity services are reliable;
- spare parts, maintenance, staff training and technical support services are easily available;
- staff can be counted upon to operate computer systems. This implies not only capacity to prepare programmes, also timely insertion into the computer of current data and elimination of what is out of date.

There remains the cost–benefit calculation. Where clerical staff are available at low cost, computer support services are unreliable, and where there is a risk that strategic information held in the computer may be inaccessible when needed, the benefits should be very clear before expensive equipment is purchased. For periodic problem-solving, computer services can be hired.

Assessing performance

Commonly accepted indicators of performance are:

profitability

return on capital employed

market standing
quality of service
innovativenes
social responsibility.

Profitability

Profitability means that at the end of the year the marketing enterprise achieves a net positive balance after allowing for all costs. How big that net balance should be will depend on the attitude of the owner. A private operator will expect a profit at least equal to what he could earn in alternative occupations open to him. If he lives in an environment of low wages and high unemployment he may accept a very low net income for lack of an alternative.

The transnational with other sources of income may be prepared to take a long-run view on profitability. This has been the attitude of Cadbury in India, for example, interested in maintaining a presence in expectation of future opportunities. More characteristic, probably, is the attitude of Gulf and Western Inc. It sold its holdings in the Dominican Republic in 1984 when low sugar prices and high taxes meant that they contributed less to its earnings than its investments elsewhere.

A marketing or input supply co-operative may be satisfied with a very small profit. Its primary concern is to serve its members by providing a favourable outlet for their produce and/or supplying inputs conveniently at low prices. An income sufficient to cover its costs will in these circumstances be judged sufficient.

For a parastatal enterprise, profitability may have still lower priority. Stabilizing the market for producers and consumers, earning foreign exchange for the national economy, and redistributing income in accordance with concepts of equity, can be prior considerations. Unless, however, the government concerned is in a position to subsidize it, a parastatal, like a co-operative, must cover its costs.

Return on capital employed

An adequate return on the capital employed in a marketing enterprise is important where the sums involved are substantial and have either been borrowed by private owners or could be used by them in some other income-earning investment. The criterion for such investors is that the return is at least as high as could be obtained from other available investment opportunities.

The risk of losing the investment must also be taken into account. If a safe government bond pays 5 per cent interest net of inflation then a private business investment should pay substantially more. For a transnational investing in a country where it faces political, exchange rate and other risks additional to those of a familiar marketing business, a return of 20 per cent on external capital invested could be a normal target rate.

In 1980, Al Haji earned only $340 on his investment of over $2000 in equipment for his enterprise as buying agent for the North Cameroun Marketing Board. His rate of return was 17 per cent; 1980 was a bad year – he could expect to earn more. Matlhaku's 1982 profit, $11 500, was a return of about 29 per cent on his equity. For these operators the return on capital includes remuneration of their own labour and management skills. Enterprises employing a manager would charge his salary as a cost and arrive at a net return on capital invested. The return on Unilever's investment in Turkey was over 20 per cent in recent years; in 1965 it was only 9 per cent (see Table 4.7).

Returns on capital employed receive less attention in parastatals. Their original capital may have been provided from public sources without obligation to pay interest. Some governments charge interest on fresh capital at a concessional rate and expect the parastatal to obtain bank finance for its purchases and stocks; they help in this by providing a guarantee. While a marketing parastatal may well be justified on service and social grounds it is appropriate, nevertheless, that it should account at market rates for the capital it employs. The advantages it offers may then be set against its full cost.

Market standing

Market standing or market share is another objective measure of an enterprise's performance. If it has a 20 per cent share of total sales of a certain product or products as against 12 per cent some years ago, then it has performed well in this regard. A declining share would imply the reverse.

This measure can be applied to most marketing enterprises including parastatal monopolies where they compete on export markets. The Cyprus Potato Marketing Board, for example, was selling more potatoes in the U.K. and at a higher price in 1984 than ten years earlier. The manager was aware, however, that competing exports from Egypt were gaining ground. For Cyprus to retain its market share it might have either to accept a lower price, or spend considerably more on promotion.

The market standing of a firm is high if its share of the market is at the higher income end, and if it has a reputation for quality produce and for reliability in its dealings. It is unwise, however, to rest on such a reputation for long. An economic recession or other adverse shift in the market may find it unable to match more innovative or cost-conscious rivals.

Quality of service

This is not easily measured, but is well understood by the clients of a marketing enterprise. Al Haji was liked by farmers because he paid them promptly, used correct measures, did not keep them waiting and was always polite. A fertilizer distributor would be providing good service if he could have ready for farmers the types and quantities of fertilizer they wanted at the time they wanted them, if he could explain clearly how they should be used and could arrange for purchases on credit.

Wimalejeewa, the commission agent, operates purely on service. His clients trust him to receive their produce, arrange for its presentation and sale at the best possible price, and to transmit the proceeds back to them. They have to trust his judgment in selling and his honesty over the price and deductions for expenses. Provision of credit to be set against the proceeds of subsequent sales was also part of his service.

Innovativeness

This means readiness and capacity to introduce new techniques, adopt new forms of organization, develop new markets. It is important if there is to be progress in marketing in a particular sector of agriculture.

New techniques are quickly taken up after they have been demonstrated as effective under the prevailing conditions. It is the role of the innovator to identify them and try them out. For Soex and the other exporters of green beans from Senegal, the use of air transport and special cartons designed to minimize its cost was a great innovation. Cassava chips became the main foreign exchange earner of Thailand within a decade of the development of new technologies and markets by European transnationals and their local partners.

Social responsibility

Consideration for the welfare of the people with which a marketing enterprise is in contact is an intangible, but also very important criterion of its performance. To a considerable degree there is a coincidence of longer-run interests. An enterprise that buys produce must treat its suppliers fairly well if it is to expect to have their business the following year. One selling supplies cannot overcharge for long or else farmers will look for another source or change their production pattern. Too soft an attitude, however, over quality control for example, is not in the interests of the growers concerned. It can mean that the eventual buyer is dissatisfied and that his business is lost. Co-operatives often have difficulties in grading. They find it hard to reject their own members' produce. The outstanding success of the Ha-ee village group in Korea reflects its leader's insistence on doing just that in order to build up a unique reputation for quality.

While they may vary prices according to quality, market conditions and terms of payment, most marketing enterprises find it advisable to offer their customers much the same terms, particularly those customers whose business constitutes the mainstay of the enterprise. They are concerned to avoid a reputation for unfair treatment, favouritism or unreliability.

A co-operative or parastatal may go further in demonstrating equity of pricing and service. It may absorb extra transport costs on produce collected from distant growers, for example. The management should recognize that this involves extra costs, however, and, in consequence, a poorer service to those producers who are more conveniently placed for the market. Thus, against a high performance rating on social responsibility, a lower rating for service, and for cost may have to be set. A good marketing manager has to be able to strike a balance between competing criteria of performance.

Issues for discussion

1. In what ways can marketing managers of co-operatives and para-statals provide effective signals to producers on market requirements? Give some illustrations.
2. Prepare, as manager of a marketing enterprise, sales plans for three agricultural products of your country (a) on domestic markets, (b) on export markets. Appraise the scope for promotion by various methods.
3. Some marketing enterprises commit up to 10 per cent of sales income to promotion of new products. Comment in this light on the sales policies of the enterprises featured in this text.
4. Propose an additional activity for one of the enterprises for which

the case study includes an income and expenditure account. Construct a set of figures for Table 8.2 demonstrating estimation of its contribution to the general overhead costs of the enterprise.

5. Derive from the data provided in the case studies some significant ratios. What do they tell you regarding the financial efficiency of the enterprise concerned?

6. Construct a probable income and expenditure statement for Kallu's enterprise (a) as a producer selling only his own produce and (b) as producer and wholesaler, on the basis of the information provided in the case study. What improvements are needed in your figures for situation (b) to make the enterprise profitable; how could they be attained?

7. On the basis of its success with potatoes the manager of the Cyprus Potato Marketing Board was asked to consider handling carrots and table grapes. Prepare an action plan for each of these products and a tentative budget along the lines of the income and expenditure statements for potatoes.

8. Assess the performance of some of the enterprises for which there are case studies; of some other enterprises with which you are familiar.

Further reading

Abbott, J. C. and J. P. Makeham, *Agricultural economics and marketing in the tropics*, Longman, London, 1984.

Baker M. J., *Marketing strategy and management*, Macmillan. London 1985.

Barry T. F., *Marketing: an integrated approach*, The Dryden Press, Chicago, 1986.

Food and Agriculture Organization, *Marketing Guides for fruit and vegetables, eggs and poultry, livestock and meat, rice, fertilizers and on overall marketing improvement. Agricultural Services Bulletins on the processing of specific products, on the organization of fruit and vegetable packing and on storage management.* F.A.O., Rome.

Foxall, G. R., *Strategic marketing management*, Croom Helm, London, 1981.

Harper, M., *The African trader: how to run a business*, East African Publishing House, Nairobi, 1973.

Harper, M. and K. Ramachandran, *Small business promotion: case studies from developing countries*, Intermediate Technology Publications, London, 1984.

Ingle M. D., N. Berge and M. Mainatton, *Microcomputers in development; a manager's guide*, Kermarien Press, West Hartford, Conn. 1984.

International Trade Centre. *Advisory manuals on export markets for specific products and on export marketing procedures; lists of potential importers.* I.T.C./G.A.T.T. Geneva.

Jones, S, F., *Marketing research for agriculture and agribusiness in developing countries: courses, training and literature*, Tropical Development and Research Institute, London, 1985.

Keegan W. J., *Maltinational marketing management*, 3rd edition, Prentice Hall International Inc., London, 1985

Kotler, P., *Marketing management: analysis, planning and control*, Prentice Hall Inc., Englewood Cliffs, N.J., 1984.

Nwokoye, N. G., *Modern marketing for Nigeria*, Macmillan, London, 1980.

Tropical Development and Research Institute. *Advisory manuals on the utilization of specific agricultural products, on processing technology and on markets for products and by-products.* T.D.R.I.. London.

Index

Accountability 152, 175–6
Accounting systems 126, 175, 189, 204–5
Accounts:
  income and expenditure 32, 40, 50, 80,
    90, 96, 107, 112, 149, 155, 160, 204
  balance sheets 50, 125, 148, 169, 204–5
Adam Smith 21
ADMARC 28
Advertisement 19, 26, 77–8, 85–9, 167,
  199–200
Agricultural factors 6, 94–5, 98, 105, 114,
  138, 155, 158–9, 212
Aid agencies 10, 135–6, 174, 180
Arbitrage 14
Assembly 185

Babcock, H. B. 9
Bananas 98, 128–32, 184–5, 189
Banks 35, 47, 49–51, 169, 194–5, 202, 210
Bargaining power 10, 24, 105, 118, 132–3,
  184
Bartels, R. 23
Bauer, P. T. 26
Branding, brands 20, 79, 162, 183–4, 200
Brokers 156–7
Budget 125, 205
Business controls 204–8

Capital 79, 94, 84–5, 98, 100, 114, 119, 134,
  158, 181–2, 188
  fixed 99, 152
  return on 205, 209–10
  working 31–2, 35, 43–7, 54–5, 121, 125,
    152, 194–5
Cash flow ratio 204
Cassava 45, 69–73, 100, 105, 208, 211
Central Soya Inc. 75
Charbonneou, J. and R. 4
Chocolate 78–84
Cocoa 78–84, 179, 186, 197, 199
Coffee 8, 46–7, 78, 123–6, 184–6, 197, 199
Collusion 25, 187
Commission sales 34–5, 43–5, 117–18, 133,
  180, 211

Commonwealth Development
  Corporation 151, 157
Comparative advantage 26
Competition 15–20, 24, 55, 66, 68–9, 83, 93
  98, 105, 117, 128, 175, 185, 189
Computers 69, 207–8
Consumer goods 46, 52, 182–4, 200
Consumers 24, 27, 108, 139–41
  preferences 6, 193–4
Contracts 3, 64–7, 72, 74–8, 94–5, 130–2,
  170, 179–80, 198
Convenience goods 24
Cooling methods 92, 130, 162, 201
Co-operatives 5, 9–10, 34, 38, 40, 45, 64–6
  91, 105–37, 139, 166, 179–80, 184–5,
  188, 194, 197, 207, 209, 212
Corn Products Inc. 184
Corruption 18, 176, 187
Costs 23, 34–8, 44, 47, 58–9, 72, 113, 118,
  125, 143, 145, 150, 174, 180–1, 186–7,
  189, 199, 202, 211
  controls 205–6
Cotton 91, 152–7, 180, 185–6
Cundiff, E. W. 24
Currency issues 46–7, 206–7
  controls 79, 101
Current ratio 204

Decision making 56, 134, 182, 188
Delivery quotas 153, 159–61
Del Monte 184
Demand 2, 13–21, 193, 196
Deng Xiaoping 25
Depreciation allowances 158, 168–9, 204,
  207
Development 2–3, 13–30, 98, 135
Disease transmission controls 162–4, 172
Dividends 79, 90, 102, 109
Drucker, P. J. 3, 23

Economies of scale 18, 55–6, 182, 186–7,
  200
Education 26, 126, 189
E.E.C. requirements 73, 162–4

Eggs 25–6
Elasticity of demand 15–16, 18, 196
  of supply 15
Employee relations 76–7, 195
Employment 31–3, 38, 40, 66, 97, 109, 157,
  164, 209
Entrepreneurs 23, 51–6
Equity 209, 211
Ethnic issues 7, 135, 138, 187
Exchange rates 27, 206
Exporting, exports 28, 31–5, 67–73, 154–6,
  162–72, 183–4
Extending activities 202–4

Facilitating services 3–4, 60–2
Family labour 57, 150, 154–5, 181
F.A.O. 22, 167
Farmers' attitudes 6–7, 9, 134–5, 188–9
Fay, R. C. 9
Feed mixing 69, 75–6, 111, 208
Fertilizers 95, 112–13, 119–22, 126–30,
  179–80, 184–5, 208, 211
Finance 31–32, 35, 49–51, 61, 68–9, 99–101,
  117, 124–5, 145–6, 168–70, 184, 188,
  197, 210
Fish 25
Forward sales 69, 197, 207
Friedman, M. 26
Fruit 25–6, 30–1, 187, 201

Galbraith, J. R. 23
Geest Industries Ltd 129–31
George, S. 93
Government departments 61, 136, 160–3
  intervention 4–5, 10–11, 22, 24, 95–7,
    112–13, 119–23, 138–9, 151–2, 159,
    175–6, 185–6, 210
  marketing services 5, 8, 60–2, 136–7, 181
  role in marketing 3–5, 25–8, 135–6, 149
Grading 19, 34, 66, 110, 153–5, 160, 166,
  212
Grain 21, 25, 27, 41–3, 45–7, 106, 186, 201
Gulf and Western, Inc. 188, 209
Gum arabic 52

Hedging 207
Helmberger, P. 9
Hides 161–2
Holton, R. 23
Hong Kong Market Authority 187

Import controls 27, 162–4
Incentives 21, 28, 55, 94–5, 127, 146
Inflation 96, 102, 138, 206–7
Initiative 56–7, 182
Innovativeness 24, 211
Inspection 47, 160, 176

Insurance 112, 207
Integration 23–4, 68–71
Interest charges 47, 66, 72, 126, 152, 158,
  210
Intermediation 103
Inventory control 121–2, 207
Investment 42–3, 164, 173–5, 184, 201–2,
  207–10

Jones, W. O. 22

Killick, T. 22
Kinship ties 51–2, 58, 115, 181, 184

Leadership 23, 119, 184–5
Legislation 3, 79, 89, 100, 134, 157
Letters of credit 72, 90, 146
Livestock 48–9, 110, 113, 158–65, 182
Location 36, 48–9, 202

Management 18, 23, 47, 49, 55, 59, 98–100,
  108–12, 126–31, 134–6, 146, 150–2,
  167, 176, 183, 188–9, 192–213
Margarine 84–93
Marginal cost 16
  return, 202–3
Mark up 198
Market 13–14, 17, 111–13
  flows 168–9, 172, 186, 196
  information 4, 20, 57, 61, 139–43, 153
  research 31, 191–4
  segmentation, separation 18–20, 186,
    193
  share, standing 117, 128, 210–11
  structure 68–9, 179
Marketing boards 11, 19, 41–3, 136, 152–7,
  165–73, 184–6
  channels 144, 158–9, 185
  efficiency 24, 55, 152, 176, 179–91,
    206–12
  infrastructure 22, 41, 60, 109, 123, 196
  margins 37, 45, 72–3, 123–8, 143, 145,
    147, 171, 197–8, 202, 206
  meaning 1–2
Marxism 25–7
Meat 25, 47–51, 182, 187
Membership issues 108–9, 114–19, 123,
  132–4, 154–5, 212
Mentzer, J. T. 22
Milk 111, 186
Milling 75, 78, 111, 146
Monopoly 16, 22, 25, 27, 128, 136, 156–8,
  165, 172–6, 185–7, 189, 196, 210
Monopsony 17–18
Myrdal, G. 26

Nason, R. W. 23

Oils edible, oilseeds 25, 52–4, 84–93, 154–6, 197
Oligopoly 18
Operating margin 198

Packing, packaging 19, 35, 66, 75–6, 88, 92, 117–18, 129–32, 153, 166–7, 194, 200, 211
Parastatals 10–11, 27, 41–3, 45, 138–80, 185–90, 194, 202, 209–12
Payment methods 161, 169
Performance criteria 128, 208–12
Planning 5, 24–6, 135, 174
Poultry 25, 74–8, 81
Price differentials 189
  discrimination 18–20, 176, 186
Pricing 13–21, 25–6, 42, 69, 73, 76, 95–6, 102–3, 105–6, 112, 124, 132, 141–3, 174, 189, 195–7, 211
Private enterprise 7, 10, 22, 28, 31–62, 104, 133, 143–6, 179–82, 185–7, 194 198–9
Processing 24, 26, 74–8, 82, 101–2, 114–18, 120, 128, 148–51, 154–5, 164–5, 170–1, 173, 179–81
Producers 24, 26, 74–8, 82, 101–2, 114–18, 126, 128, 148–151, 154–5, 164–5, 170–1, 173
Profit, profitability 10, 19, 24, 27, 35, 38, 46–7, 65–6, 90, 97, 100–2, 173, 209
Pulses 21, 98, 186
Purchasing methods 91–2, 120, 142–3, 153–4, 197–8

Quality 20, 32, 38, 40, 42, 72, 100, 106, 122, 186, 197, 211
  control 32, 100, 117–18, 126, 130–2, 147, 150, 154–7, 212

Refrigeration 34, 161, 201
Regulatory services 3–4
Representation of producers 123, 151, 168, 173, 176
Research 93, 129–32, 156
Retailers, retailing 8, 23, 37, 45–51
Rice 39–41, 110–14, 138–46
Risk 23, 28, 73, 133–4, 183, 198, 207, 210
Rostow, W. 23

Sacks 41–3, 139
Sales methods 92, 117–19, 162–3, 166–8, 171, 198–9
  promotion 78, 167, 183–4, 199–200, 210

Salaries 134
Samli, A. C. 22
Scitovsky, T. 26
Seasonality 31, 38, 41, 116–17, 161, 201, 204
Seeds 81–2, 91–2, 101–2, 152–6, 160, 179, 184
Site of operations 195–6, 202
Shipping 68–9, 72, 166
Slater, C. 23
Social marketing 24, 139, 211–12
Sorghum 41–3, 52
Sorting 117, 126
Spinks, R. 5
Stabilization 11, 41–3, 123–7, 138–43, 165, 170, 172–6, 185–6, 189, 209
Staffing 176, 181, 189, 195, 208
Standards, standardization 23, 98–100, 117–18, 132–3, 172, 186
Stocks 92, 120–3, 125, 139–43, 172, 185–6
Storage 14, 23, 106, 110, 146, 201–2, 207
Subsidies 111–13, 121, 128, 143–5, 186, 189, 209
Sugar 8, 25, 93–8, 209
Supply 13–21, 121, 162–3, 167, 177, 212

Tannery 162–4
Taxes 22, 79, 83, 89, 102, 111, 134
Tea 8, 25, 98, 146–53, 173, 199
Technology 99–100, 183–4, 188
Tobacco 26, 103, 180
Training 2, 5, 43, 61–2, 65–7, 72, 99–101, 110, 114, 123, 126, 132, 136, 167–8, 200–1, 211–12
Transnationals 6–8, 21–2, 65–103, 179–80, 183–4, 186, 194, 209, 211

Vegetables 25–6, 30–8, 43–5, 107–19, 165–72, 182, 187
Vent for surplus 21–2

White, P. O. 23
Wholesalers, wholesaling 7, 31–47, 51–6, 82–3, 92, 117–18, 166, 197
Women in marketing 7, 31–3, 49
Working hours 35, 38, 48, 56–7, 88, 111, 182
World Bank 103, 123

Yams 106